MACHINE DESIGN I

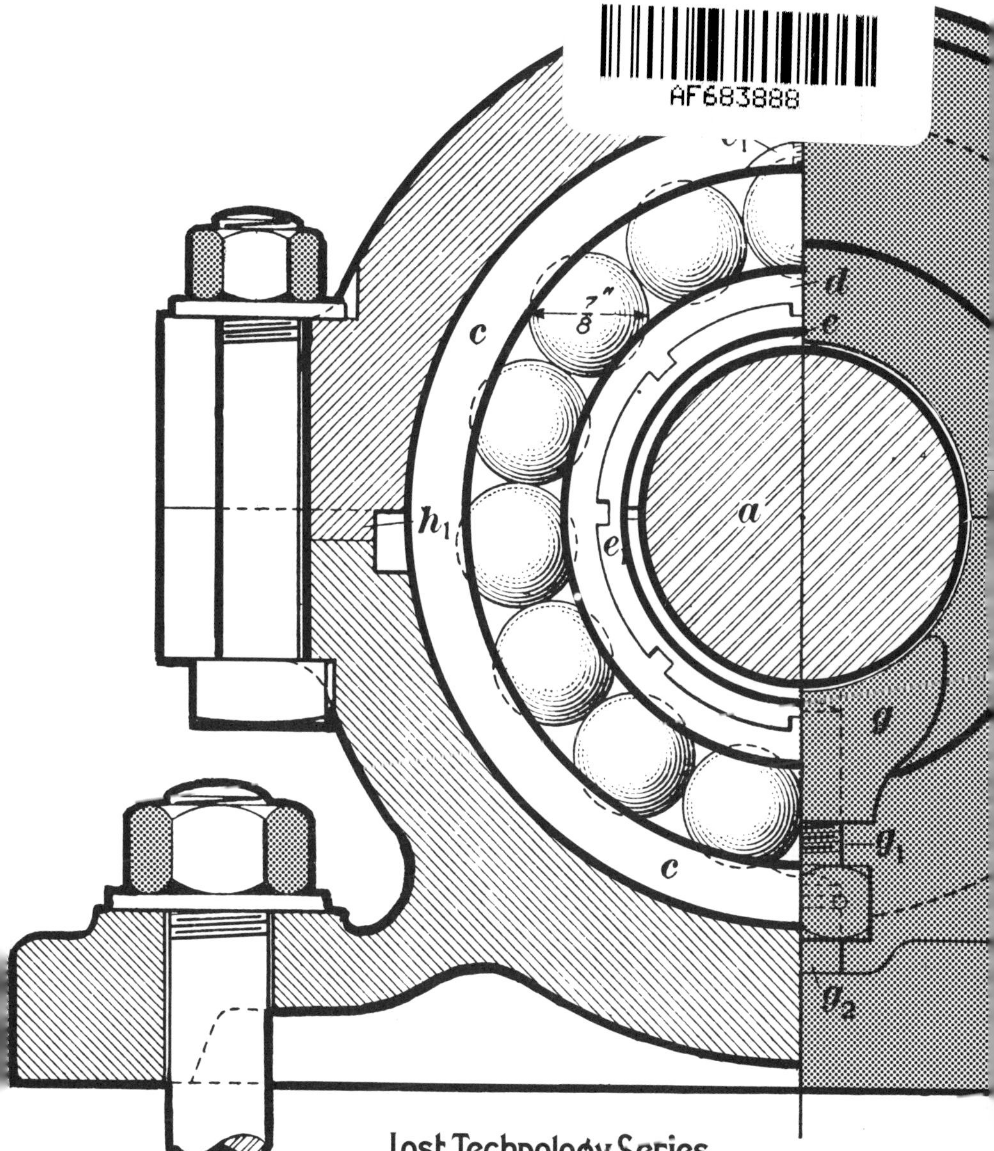

Lost Technology Series

Reprinted by Lindsay Publications Inc.

Machine Design I

I. C. S. Reference Library

ISBN 0-917914-33-3

2 3 4 5 6 7 8 9 0

I. C. S.
REFERENCE LIBRARY

A SERIES OF TEXTBOOKS PREPARED FOR THE STUDENTS OF THE INTERNATIONAL CORRESPONDENCE SCHOOLS AND CONTAINING IN PERMANENT FORM THE INSTRUCTION PAPERS, EXAMINATION QUESTIONS, AND KEYS USED IN THEIR VARIOUS COURSES

LINK MECHANISMS
GEARING
GEAR TRAINS AND CAMS
PULLEYS AND BELTING
MATERIALS OF CONSTRUCTION
STRENGTH OF MATERIALS
THE TESTING OF MATERIALS
MACHINE DESIGN

SCRANTON
INTERNATIONAL TEXTBOOK COMPANY
172

WARNING

Remember that the materials and methods described here are from another era. Workers were less safety conscious then, and some methods may be downright dangerous. Be careful! Use good solid judgement in your work, and think ahead. Lindsay Publications, Inc. has not tested these methods and materials and does not endorse them. Our job is merely to pass along to you information from another era. Safety is your responsibility.

Write for a catalog or other unusual books available from:

Lindsay Publications, Inc.
PO Box 12
Bradley, IL 60915-0012

MACHINE DESIGN

(PART 1)

DESIGN IN GENERAL

DESIGN OF DETAILS

1. In this Section, rules and formulas for designing many of the most common details of machines are given. These rules are in some cases based on considerations of strength, as treated in *Strength of Materials;* in other cases, the wear to which the parts are to be subjected has been the principal element in determining the given proportions. In all cases, however, the practice of successful designers has been followed in preference to mere theoretical principles.

2. The first work a young designer is called on to do is usually that of making drawings of details of machines, the general plans of which have been developed by his superiors. The principal dimensions of these details will be given him, and he will be required to furnish drawings from which patternmakers, blacksmiths, and machinists can make up the parts ready to be put in their proper place in the machine.

3. In most shops, such parts as bolts, nuts, screws, pipe fittings, etc., and often other simple parts of machines, are obtained from factories, where they are made in large quantities by special machinery. If the shop is a large one, there may be a separate department where these parts are made according to fixed standards. The designer should ascertain the practice of the shop in this regard, and in all cases make the details conform with these standards. He should also know the kind of material available, the methods employed by the shop in working this material, and the capacity and principal dimensions of the tools and appliances for doing the work, in order that the details as designed may be built in the most economical manner. When designing a machine part, the processes that will be used in making it should be kept in mind; this will often prevent constructions that would be very difficult and expensive if built with the machinery in use in the shop for which the design is made.

4. In making designs of details, it is always advisable to draw them to as large a scale as can be used conveniently. The scale commonly used for small details is full size; for larger ones, 6, 3, or $1\frac{1}{2}$ inches to the foot may be used. A scale of 4 or 2 inches to the foot should never be used if it can be avoided.

5. It should be remembered that the object of a detail drawing of a machine part is to show the workmen in the clearest possible manner how the part is to be made and finished, so that it will take its proper place in the completed machine and do the work for which it is intended. The designer must therefore be very careful to make the drawing show the form and dimensions of each portion as clearly as possible; the drawing should also show plainly the kind and quality of material to be used, and the finish, if any, to be given the different surfaces. Sections should be used wherever the general views do not show the form with perfect clearness.

It is well for the designer to imagine himself in the position of a man in the shop who knows nothing of the machine, and to study his drawing carefully to see whether anything can possibly be lacking that will be required to make perfectly clear the ideas he wants carried out.

6. Ordinary dimensions are expressed in feet and inches, and the fractions $\frac{1}{2}$, $\frac{1}{4}$, $\frac{1}{8}$, $\frac{1}{16}$, etc. of an inch. The fractions $\frac{1}{3}$, $\frac{1}{5}$, or $\frac{1}{6}$ should never be used in dimensions, as the scales that mechanics use are not divided into these fractions. The most common scales in use by mechanics for ordinary work are 2-foot rules divided into inches, numbering from 1 to 24; for this reason many draftsmen give all dimensions less than 2 feet in inches and fractions of an inch, and dimensions greater than 2 feet in feet, inches, and fractions of an inch.

Unless great accuracy is required, decimals are never used in giving dimensions. If decimal values are obtained from the calculations, they are expressed in the nearest $\frac{1}{8}$, $\frac{1}{16}$, or $\frac{1}{32}$ inch, according to the degree of accuracy required. In particular cases, where extreme accuracy is needed, the dimensions may be expressed in decimals; as, for example, in denoting the pitch of gear-teeth.

GENERAL DESIGN OF A MACHINE

7. The methods to be employed in the general design of a machine will vary so much with different conditions that no fixed rules or methods of procedure can be given. The first thing necessary is a thorough knowledge of the work that the machine must do, together with its location and surroundings and the conditions under which it must perform its work. Keeping these facts in mind, the designer must apply his knowledge of the principles of mechanics, strength of materials, and the design and construction of details in such a way as to accomplish the desired end in the simplest and most direct manner consistent with the conditions imposed.

All machines consist of different combinations of simple principles, and in order to be successful, the designer must become thoroughly acquainted with these principles and the relation they bear to one another. A study of machines that have been built for similar work will greatly assist in suggesting ideas for the new machine.

In many cases, it will be necessary to make more or less complete drawings of a number of different plans before a satisfactory result can be obtained. A combination that appears feasible at first may be found to be impracticable when drawn out in detail and the parts proportioned so as to give the necessary strength. In other cases, the motion of some part may be found to be limited in such a way as to interfere with the proper working of the machine. The difficulty or expense of manufacture may also make some otherwise good design impracticable.

8. In all classes of work—whether it is the designing of details or of more complicated combinations—all calculations, notes, and sketches should be kept in such form that they may readily be preserved for future reference. These notes should be dated and given titles that will make their purpose perfectly clear. In this way, ideas that may be impracticable for the particular case for which they were originally developed can be kept for a possible future use; moreover, the results of many hours spent in calculation will be preserved and thus make repetition unnecessary. Some engineering establishments supply their draftsmen with books made of Manila paper bound in board covers, and all calculations, notes, and sketches are made in these books instead of on loose sheets of paper that are likely to become lost.

9. The following practical rules are often neglected by inexperienced designers:

Make all parts that are subject to wear or breakage accessible for the purpose of inspection, repairs, or renewal.

Provide means for adjusting all parts that are subject to wear.

Make careful provision for lubrication.

Use links and rotating pieces for guiding motion in preference to slides.

Use cranks, levers, belts, and gear-wheels for transmitting motion in preference to cams, screws, or worm-wheels.

Wherever possible, make the motion of all parts positive; that is, avoid the use of weights or springs for producing motion.

Use through bolts or **T**-*headed bolts instead of tap bolts or studs, wherever it can be done.*

10. Designers are often required to furnish an estimate of the weight and cost of a machine from the drawings.

This is done in the following manner: The volume of the different details is estimated by means of the principles of mensuration; the weight can then be obtained by multiplying the volume of each piece by the weight of a cubic unit of the material of which it is composed. When the weights are known, the cost of the material is easily found from the known market values. The time that will be required for fitting and finishing the different pieces is then estimated and charged for according to the rates paid for that work. In this way the cost of the machine may be estimated with a degree of accuracy that will depend on the knowledge the estimator has of the time required to do different kinds of work in the shops, and his skill in making approximate calculations of the volumes of irregular-shaped bodies.

PROPERTIES OF MATERIALS

11. Cast Iron.—The general properties of the various metals used in the construction of machine parts have been given in *Materials of Construction*, but some additional remarks about cast iron are necessary.

The great advantage of cast iron is the ease with which it may be given any desired form. Shapes that cannot possibly be forged from wrought iron may be cast with comparative ease. The operation of casting is as follows: A pattern having the exact shape of the required part is first made. This pattern is usually of pine, though metal is sometimes employed when the castings are small and a great number are required. The pattern is placed in a bed of sand or loam, in which it leaves, after being removed, an impression, or cavity, called the *mold*. Molten metal is then poured into the mold, and, after cooling, the casting is withdrawn and finished to the required dimensions.

In cooling, cast iron contracts about $\frac{1}{8}$ *inch per foot in each direction*. This contraction, however, may vary from $\frac{1}{16}$ to $\frac{3}{16}$ inch per foot, according to the condition of the iron used and its preliminary treatment. On account of this contraction, commonly called the *shrinkage*, the pattern must be made larger than the required casting. Therefore, in practice, a *shrink rule*, which is about $\frac{1}{8}$ inch longer per foot than the standard rule, is always used in constructing the pattern.

12. A serious difficulty experienced in the use of cast iron is its liability to be thrown into a state of *internal stress*, on account of inequality of cooling after being poured into the mold. It is a matter of experience that the amount of contraction depends on the size and thickness of the casting. In general, thick and heavy parts contract more than thin ones; consequently, a casting composed of both thick and thin parts will sometimes differ from the form desired. Again, one part of a casting may cool and solidify while another part is still in a molten condition; therefore, the contraction of the latter must strain the part already solidified. Internal stress is thus produced, which must to some degree reduce the effective strength of the casting.

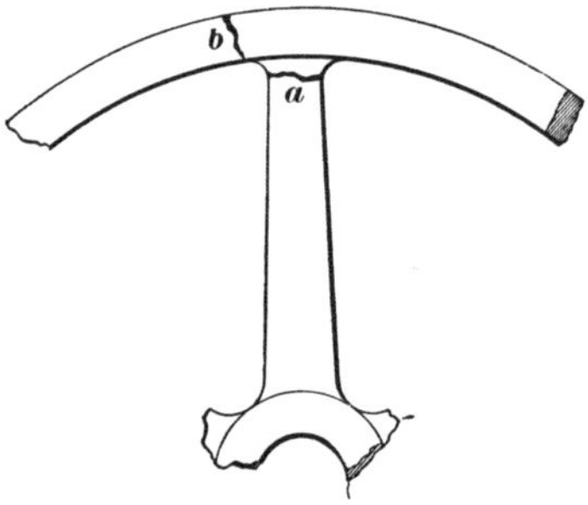

FIG. 1

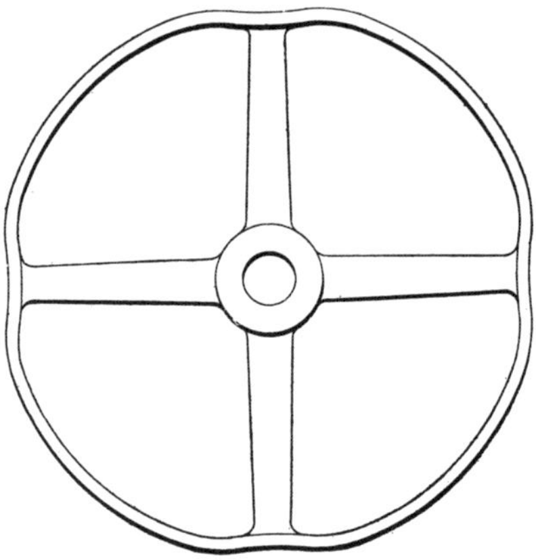
FIG. 2

Take, for example, the case of a pulley. When the rim is thin, but rigid, it is likely to contract and solidify first, and the subsequent contraction of the arm may induce a fracture, as shown at *a*, Fig. 1. If, however, the arms set first, the subsequent contraction of the rim may cause a fracture, as shown at *b*.

13. When pulleys are cast with thin rims that are not rigid, the casting sometimes takes the form shown in Fig. 2. The rim is drawn in at the points where it joins the arms, because the arms solidify and contract after the rim has set, and the latter, not being sufficiently rigid to withstand the pull of the arms, is distorted.

These internal stresses make cast iron an unreliable material for the construction of parts requiring strength; and it should be the aim of the designer to prevent these stresses, as far as may be, by making all parts that are to be cast as uniform in thickness as possible, and avoid having a large boss or hub appear in a comparatively thin part, or having a very thick part meet a very thin one.

14. It is found that, in cooling, the iron crystals arrange themselves perpendicularly to the surface of the casting. For this reason, an inside angle m, Fig. 3 (a), is a source of

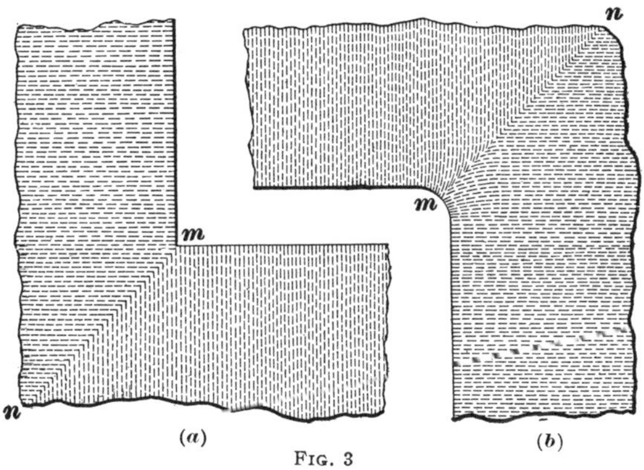

FIG. 3

weakness, the casting having a tendency to break through the line mn. Such corners should be rounded, as shown at m, Fig. 3 (b), in which case the crystalline arrangement renders the casting much stronger.

15. Variation of the Elastic Limit.—Experiments have indicated that if a machine part is subjected to repeated tensile stresses, the material shows an *increase* in its elastic limit while under a tensile stress, but that the elastic limit in regard to *compressive* stresses has been *decreased*. For parts subjected to stresses of this nature, it would evidently be unsafe to apply working stresses that approach the elastic limit.

A machine part is generally supposed to be safe if so proportioned as to withstand the *maximum* stresses to which it may be subjected. This supposition leaves out of consideration whether these stresses are constant or whether they change in value, in direction, or in both. That this view is erroneous has been shown by a large number of experiments performed by various investigators of high repute. These experiments prove that the ultimate strength of a material is *lowered* when the load varies in amount alone or in both amount and direction.

16. Maximum and Minimum Stresses.—With a constant load, the usual custom is to make the working stress of the material $\frac{1}{4}$ or $\frac{1}{5}$ of its ultimate strength, in order to allow for possible defects and unforeseen stresses. The factors of safety are then 4 and 5, respectively. To increase this factor to 6 and 7 for variable stresses, as is customary, will not always bring the working stresses within safe limits. The reason for this will be seen from the following explanations. In a revolving shaft with a variable load, the stress at any one place will change between a maximum and a minimum if the shaft is subjected to torsion only. Denoting the values of the maximum and minimum stresses by S_{max} and S_{min}, respectively, the total variation v will be $S_{max} - S_{min}$.

In a revolving loaded axle, as, for instance, on a railroad car, the stresses at any one place are constantly alternating between tension and compression. Considering the tensile stresses as positive, and the other stresses as negative, and subtracting the latter from the former,

$$v = S_{max} - (-S_{min}) = S_{max} + S_{min}$$

To show the difference between these two cases, let it be supposed that in a revolving shaft the unit torsional stress, that is, the unit stress due to the action of a torsional force, varies between 10,000 and 3,000 pounds; v is therefore

10,000 – 3,000 = 7,000 pounds. In an axle revolving under the conditions just mentioned, the stress may change from 7,000 pounds tensile to 7,000 pounds compressive. The variation v is then 7,000 – (– 7,000) = 14,000 pounds.

17. Repeated, or Periodic, Stresses.—When a material is subjected to regularly repeated stresses, its ultimate strength suffers a further reduction. To ascertain the relative strengths of various materials working under such conditions, it is assumed that if two test pieces having identical dimensions will withstand the same amount of variation in stress, repeated a similar number of times, they are of the same strength. It was found by experiment that two bars of wrought iron were able to withstand the same number of repeated stresses before breaking, when the stresses were of the following nature: In one case the stresses were tensile, varying from 31 to 18,713 pounds per square inch; in the other they alternated from a tensile stress of 8,317 pounds to a compressive stress of 8,317 pounds. The average variation in both cases was about 17,000 pounds. Assuming the ultimate tensile strength of wrought iron to be about 50,000 pounds, it is seen at once that the effect of such stresses is to reduce the tensile strength in one case to 18,713 pounds and in another to 8,317 pounds.

18. Classification of Stresses.—The various stresses to which a machine part may be subjected can be arranged in three classes as follows:

1. Stresses that do not vary.
2. Stresses that vary in value, but not in direction.
3. Stresses that vary both in value and in direction.

It has been found that the ultimate stresses that machine parts coming under these three classes may resist, are, in round numbers, in the proportions of 3 : 2 : 1. In cases where the stresses vary between zero and the maximum or where the stresses change direction, it is supposed that the changes are gradual. If the load is applied suddenly, but without velocity, the values found from the proportion just given should be divided by 2.

SCREWS, BOLTS, AND NUTS

SCREWS AND SCREW THREADS

19. Screws are used in machine construction for three purposes: (1) As fastenings for clamping or joining parts together; (2) for the transmission of motion; and (3) for producing pressure.

A screw provided with a head and a nut for fastening parts together is generally called a *bolt*.

20. Sharp V Thread.—Screw threads are usually triangular or square in section, the triangular form being best for bolts, and the square form best for screws transmitting motion. The fundamental **V** thread is shown in Fig. 4, and is termed the **sharp V thread** to distinguish it from other triangular threads in which the top and bottom of the thread are flattened or rounded off.

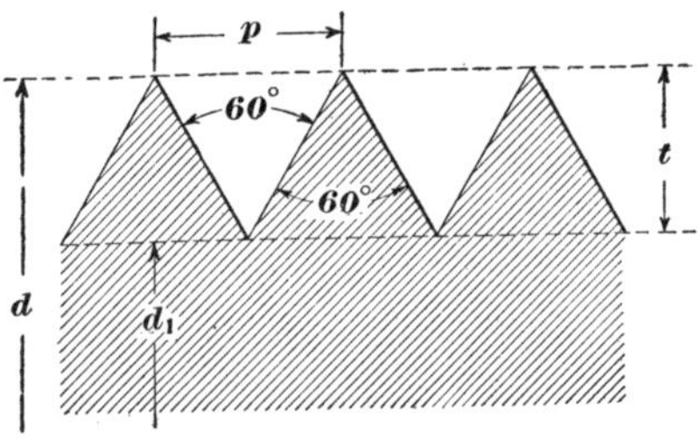

Fig. 4

The angle between the sides of the thread is 60°. The distance p from a point on one thread to a corresponding point on the next is called the **pitch** of the screw. The **lead** of a screw is the distance that a nut will move, in an axial direction, during one complete turn of the screw. In the case of a single-threaded screw, like that shown in Fig. 5, the pitch and the lead are equal.

As shown in Fig. 4, a section of a single thread is an equilateral triangle, the altitude of which is t. Hence,

$$t = p \cos 30° = .866p \qquad (1)$$

Let d = outside diameter of bolt;

d_1 = diameter at root of thread.

Then, $d_1 = d - 2t = d - 1.732p$ (2)

The diameter d_1 must always be used in calculating the strength of a bolt.

Letting n represent the number of threads per inch in a screw, then

$$p = \frac{1}{n}, \text{ and } n = \frac{1}{p} \qquad (3)$$

The pitch of the sharp **V** thread corresponds in general with that of the United States standard thread.

21. United States Standard Thread.—In 1864 an improvement on the sharp **V** thread was proposed by William Sellers and subsequently adopted by the United States Navy and War Departments. From its originator, this thread is known as the **Seller's thread,** but is more commonly called the **American thread,** or the **United States standard thread.** In Fig. 5 is shown a general view, and in Fig. 6 an enlarged section of this thread.

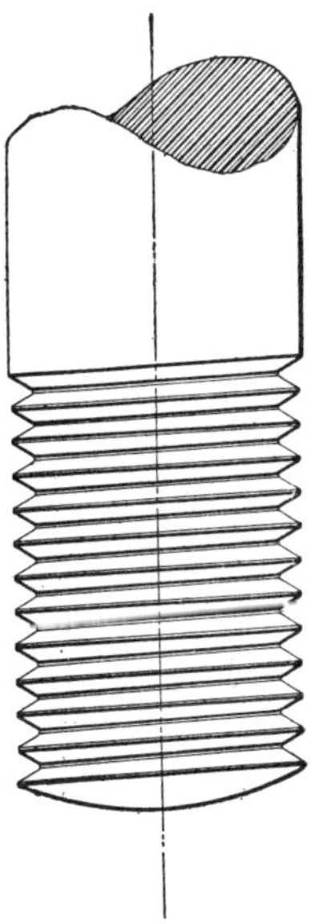

FIG. 5

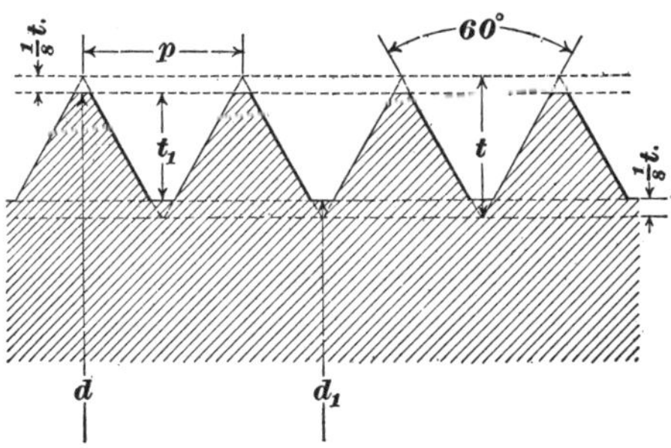

FIG. 6

The difference between the sharp **V** thread and the United States standard thread will be seen by comparing Figs. 4 and 6. Thus, in the latter thread one-eighth the altitude of the triangle is cut off from the apex, and the angle at the root is filled in to a like depth. Hence, the *real depth* of the thread t_1 is three-fourths the altitude of the triangle; that is, $t_1 = \frac{3}{4} t$.

But, $$t = p \cos 30° = .866 p;$$
hence, $$t_1 = \frac{3}{4} t = .65 p$$

22. The pitch of the United States standard thread depends on the diameter of the bolt, and may be ascertained approximately by the following empirical formula, in which d represents the outside diameter of the bolt:

$$p = .24 \sqrt{d + .625} - .175 \text{ inch} \qquad (1)$$

The diameter d_1 at the root of the thread may be found by the formula

$$d_1 = d - 2 t_1 = d - 1.3 p \qquad (2)$$

Let n = number of threads per inch;
p = pitch.

Then, $$p = \frac{1}{n}, \text{ and } n = \frac{1}{p} \qquad (3)$$

Consequently, $$d_1 = d - \frac{1.3}{n} \qquad (4)$$

EXAMPLE.—The external diameter of a bolt is $1\frac{3}{8}$ inches. Find the pitch, the number of threads per inch, the depth of thread, and the diameter of bolt at root of a United States standard thread.

SOLUTION.—From formula **1**, just given,

$$p = .24 \sqrt{1.375 + .625} - .175 \text{ in.} = .164 \text{ in. Ans.}$$

From formula **3**,

$$n = \frac{1}{.164} = 6, \text{ nearly}$$

Use six threads per inch; then, $p = \frac{1}{6} = .167$ in. Ans.
$t_1 = .65 p = .65 \times .167 = \frac{7}{64}$ in., nearly. Ans.
$d_1 = d - 2 t_1 = 1\frac{3}{8} - \frac{7}{32} = 1\frac{5}{32}$ in. Ans.

Table II, at the end of this Section, gives the number of threads per inch, diameter of bolt at root of thread, and effective area of bolt at root of thread, United States standard sizes.

23. Whitworth Thread.—In England, the **Whitworth system** of threads is used. This thread differs from the United States standard mainly in the size of the angle between the sides of the thread and in the rounding off of its point and bottom. As shown in Fig. 7, the angle between the sides is equal to 55°, and the amount cut away from the top and bottom is $\frac{1}{6}t$. The remaining corners are not left sharp, as in the United States standard thread, but are

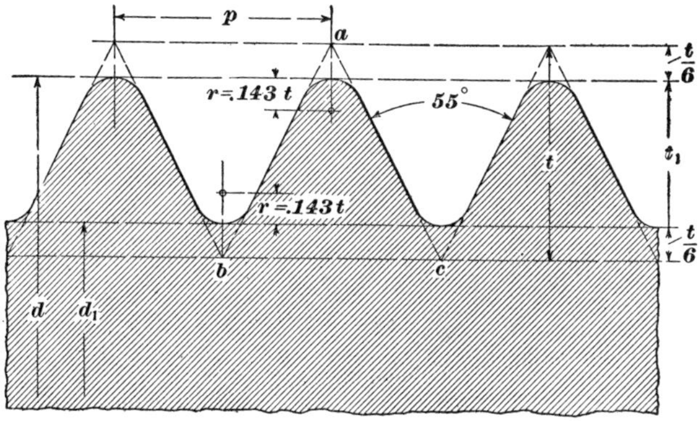

FIG. 7

rounded by an arc with a radius equal to $.143\,t$. The real depth of the thread t_1 is two-thirds the altitude of the triangle abc; that is, $t_1 = \frac{2}{3}t$.

But, $$t = .5\,p \cot 27.5° = .96\,p \qquad (1)$$
and $$t_1 = \tfrac{2}{3} \times .96\,p = .64\,p \qquad (2)$$

24. The pitch of the Whitworth thread depends on the diameter of the bolt, and may be obtained approximately by means of the following formula, in which d represents the diameter of the bolt:

$$p = \frac{1}{n} = .08\,d + .04 \qquad (1)$$

To find the diameter d_1 at the root of the thread, the following formula is used:

$$d_1 = d - 2\,t_1 = d - 1.28\,p \qquad (2)$$

The diameter d_1 must always be used in calculating the strength of the bolt.

Letting n represent the number of threads per inch in a screw,

$$n = \frac{1}{p} \qquad (3)$$

Consequently, $$d_1 = d - \frac{1.28}{n} \qquad (4)$$

EXAMPLE.—The external diameter of a bolt is $1\frac{1}{2}$ inches. Assuming that the bolt is threaded according to the Whitworth system, find the pitch, the number of threads per inch, the depth of the thread, and the diameter of bolt at root of thread.

SOLUTION.—By formula **1**,

$$p = .08 \times 1.5 + .04 = .16 \text{ in. Ans.}$$

By formula **3**, $$n = \frac{1}{.16} = 6, \text{ nearly}$$

Using six threads per inch, $p = .167$ in. Ans.

By formula **2**, Art. **23**,

$$t_1 = .64 \times .167 = \tfrac{7}{64} \text{ in., nearly. Ans.}$$

By formula **2**, $d_1 = 1\frac{1}{2} - \frac{7}{32} = 1\frac{9}{32}$ in. Ans.

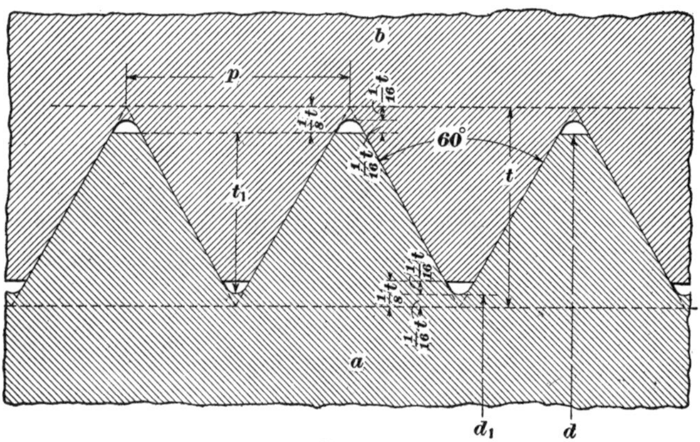

FIG. 8

25. International Standard Screw Thread.—The Whitworth system has been used extensively on the continent of Europe, but on account of the inconvenience of combining the British with the metric units of length, attempts were made at various times to adopt a more convenient system as a standard. Finally, at a meeting held at Zurich, in 1898, an international committee agreed on proportions that in the main are the same as those of the French standard screw thread. The shape of the international thread is almost identical with the United States standard thread, but it has an advantage over the latter in that a clearance has been provided at the top of the thread, both in the nut and on the bolt. As shown in Fig. 8, where *a* represents the bolt and b the nut, one-eighth of the altitude of the triangle has been cut off from the apex of the thread. The angle at the root is filled in to a depth of one-sixteenth of the total altitude, but the flat part thus produced is preferably rounded off as indicated. The real depth of the thread t_1 is therefore thirteen-sixteenths of the altitude of the triangle; that is, $t_1 = \frac{13}{16} t$. As

$$t = p \cos 30° = .866 p$$
$$t_1 = \tfrac{13}{16} \times .866 p = .7036 p \qquad (1)$$

The diameter d_1 at the root of the thread may be found by the following formula, which may be used for values given in either millimeters or inches.

$$\begin{aligned} d_1 &= d - 2 \times \tfrac{13}{16} t \\ &= d - 1.625 t \\ &= d - 1.407 p \qquad (2) \end{aligned}$$

Table III, at the end of this Section, gives the proportions of the international standard screw thread in millimeters. To facilitate comparison with the United States standard thread, however, some of the values are expressed in inches; likewise, the number of threads per inch corresponding to the pitch are given in millimeters.

FIG. 9

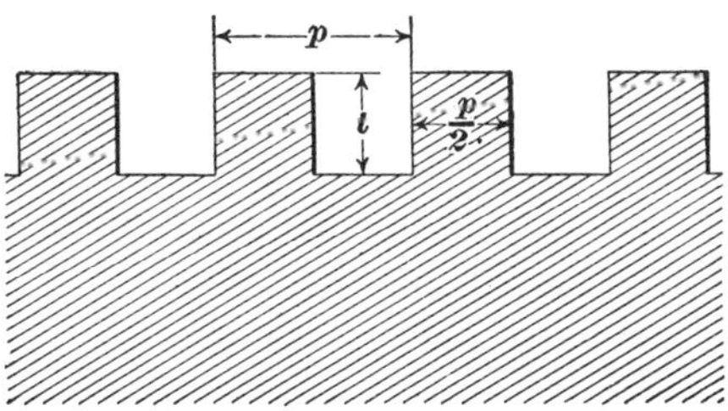

FIG. 10

26. Square Thread.—In Fig. 9 is shown a screw with a **square thread,** and in Fig. 10 an enlarged section of the thread. As the name implies, and as seen from Fig. 10, the section of the thread is a square, each side of which is one-half the pitch.

The pitch of the square thread is usually taken as *double* that of the United States standard thread for the same diameter of bolt; for example, the pitch of a square thread on a $2\frac{1}{2}$-inch bolt or rod is $\frac{1}{2}$ inch. Hence, the pitch p may be found by inserting the values given for n in Table I in the following formulas:

$$p = \frac{2}{n} \qquad \textbf{(1)}$$

and

$$t = \frac{p}{2} \qquad \textbf{(2)}$$

EXAMPLE.—Find the pitch and the depth of the thread on a $1\frac{3}{8}$-inch bolt.

SOLUTION.—From Table I, $n = 6$. Inserting this value in formula **1**, $p = \frac{2}{6} = \frac{1}{3}$ in. Ans.

By formula **2**, $t = \frac{p}{2} = \frac{1}{6}$. Ans.

The edges of the threads should be rounded off very slightly, so as to prevent them from being accidentally flattened, which would cause a nut to bind on the thread of the screw.

Several modifications of the square thread are in use. In some, the depth t is less than $\frac{p}{2}$, and in others the width of the thread is greater than $\frac{p}{2}$.

27. Acme Standard Thread.—As difficulties are encountered in making square-threaded screws and in giving the threads a smooth bearing surface, the tendency is to substitute in their place screws having threads that taper slightly from root to point, but retain the height of the corresponding square thread. By giving a taper to the thread, it does not have to be cut in the lathe, as is the case with the square thread, for it may also be cut with a die. Another advantage in favor of the tapering thread is that the nut—with which the screw engages—when composed of two parts, may be made to engage or disengage readily with the screw.

In the taper thread in general use the angle included between the sides of the thread is 29°. This variation of the square thread is known as the **Acme standard screw thread.** The angle of this thread is similar to that of the standard worm-thread, shown further on, but the threads differ in height.

In the Acme standard thread, the relation between the depth and the pitch is the same as that of the square thread; that is, the depth is one-half the pitch. To insure sufficient clearance, the bottom of the thread is lowered $\frac{1}{100}$ inch, making the true height of the thread

$$t = \frac{p}{2} + .01$$

Similarly, the diameter at the root of the thread in the nut is increased by $\frac{1}{100}$ inch, as shown in Fig. 11. The outside diameter of the screw and the diameter of the hole in the nut are not changed. This clearance of $\frac{1}{100}$ inch is constant for all sizes of screws made with the Acme standard thread.

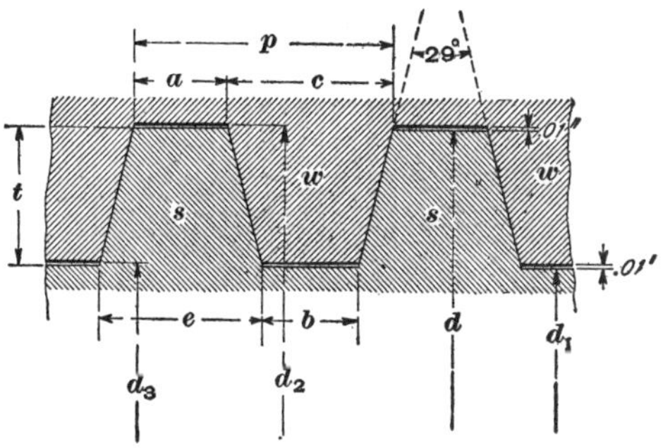

FIG. 11

28. Fig. 11 shows portions of a section through a screw s and nut w, d being the outside diameter of the bolt, and n the number of threads per inch of length. The dimensions of the various parts of the thread are as follows:

Diameter of screw at root of threads,

$$d_1 = d - \left(\frac{1}{n} + .02\right) \qquad \textbf{(1)}$$

Diameter at root of thread in nut, or the outside diameter of the tap,

$$d_2 = d + .02 \qquad (2)$$

Diameter of drill for hole in nut,

$$d_3 = d - \frac{1}{n} \qquad (3)$$

Depth of thread,

$$t = \frac{1}{2n} + .01 \qquad (4)$$

Thickness at top of thread,

$$a = \frac{.3707}{n} \qquad (5)$$

Width of space at bottom of thread,

$$b = \frac{.3707}{n} - .0052 \qquad (6)$$

Width of space at top of thread,

$$c = \frac{1}{n} - a \qquad (7)$$

Thickness at root of thread,

$$e = \frac{1}{n} - b \qquad (8)$$

Table IV, at the end of this Section, gives the dimensions of threads varying in number from one to ten per linear inch. The diameters of the screws for the various numbers of threads per inch are not given, as there is no definite relation between pitch and diameter in the Acme standard thread. The pitch and diameter of the screw will depend on the purpose for which the screw is intended.

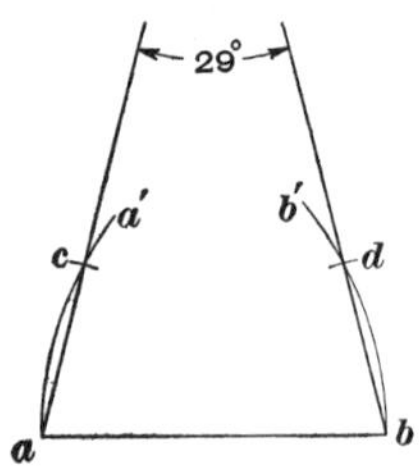

FIG. 12

To facilitate the laying off of an angle of 29°, the construction shown in Fig. 12 may be used. With the line $a\,b$ as a radius and the points a and b as centers, draw arcs $b\,b'$ and $a\,a'$. With a as a center and one-half of $a\,b$ as a radius, draw a short arc intersecting the arc $a\,a'$ at c; locate the point d in the same manner. Lines drawn through the points a and c and b and d will include an angle of 29°.

29. Knuckle Thread.—Another variation of the square thread is the **knuckle thread,** a section of which is shown in Fig. 13. In this type, the top and the root of the thread are rounded off, thus not only increasing its strength but also its friction surface. The knuckle thread is especially adapted to withstand rough usage.

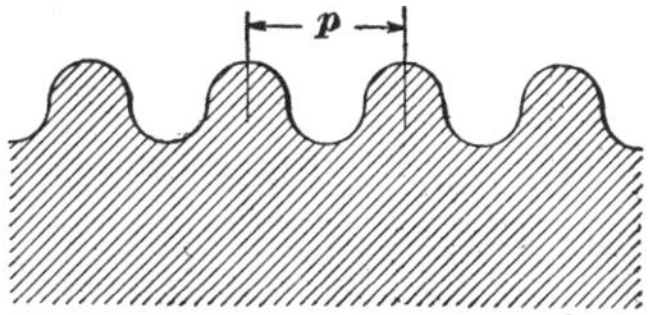

FIG. 13

30. Trapezoidal, or Buttress, Screw Thread.—As shown in Fig. 14, one face of the trapezoidal thread is perpendicular to the axis of the screw, and the other is inclined at an angle of 45°. From the construction of the figure, it is evident that $t = p$, the pitch. To form the actual thread, an amount equal to $\frac{1}{8}\,t$ is cut off from the top and bottom of the triangle; hence, the real depth t_1 is $\frac{3}{4}\,t$.

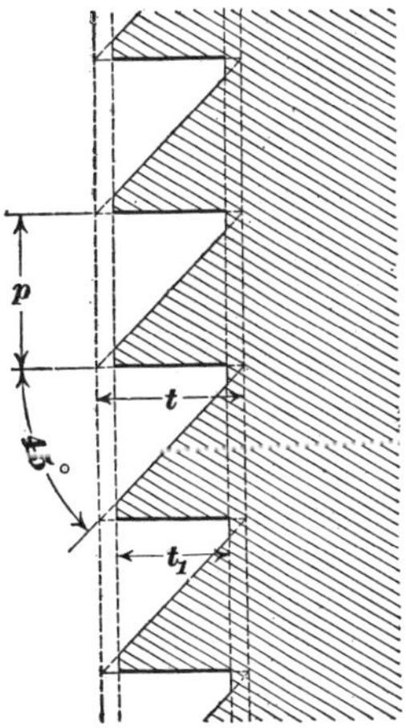

FIG. 14

When this thread is used, it is generally for communicating motion, or where great resistance without any bursting tendency of the nut is required. The usual dimensions are given by the following formulas, in which the letters have the same meaning as before:

$$p = \frac{2\,d}{15} = \frac{d_1}{6} \qquad (1)$$

$$t_1 = \frac{d}{10} = \frac{d_1}{8} \qquad (2)$$

For example, suppose that the diameter of the screw is $1\frac{1}{2}$ inches; then, the pitch of a trapezoidal thread is

$$p = \frac{2\,d}{15} = \frac{2 \times 1\frac{1}{2}}{15} = \tfrac{1}{5} \text{ inch}$$

The number of threads per inch is $\frac{1}{p} = \frac{1}{\frac{1}{5}} = 5$, and the depth of the thread is

$$t_1 = \frac{d}{10} = \frac{1\frac{1}{2}}{10} = .15 \text{ inch}$$

A variation of the buttress thread is used for breech blocks in modern guns and for bolts that fasten the armor plates to the hull of a warship. The thread is rounded off at the top and root, and the apex angle, instead of having one side coincide with a normal to the axis of the screw, is increased to 60° by letting this side make an angle of 15° with the normal.

31. Comparison of Various Threads.—The relative advantages of the various forms of screw threads may be shown by considering the forces acting on the thread.

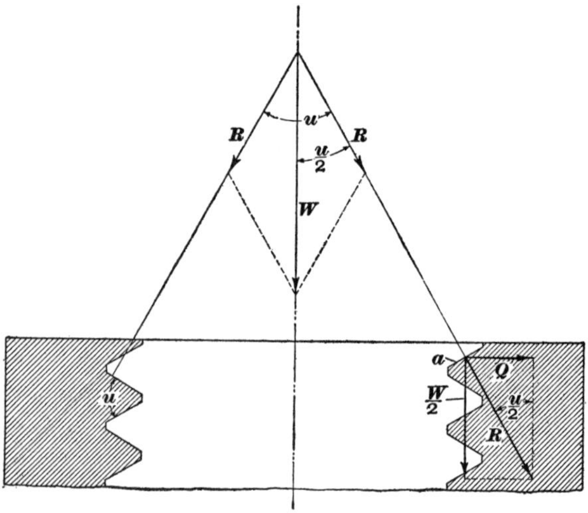

Fig. 15

Usually, the load on a bolt or screw acts in the direction of its axis; that is, a bolt used as a fastening is in tension, while a screw used to produce pressure is in compression. In either case, the load is carried by the reaction between the thread surfaces of the screw and the nut.

Let Fig. 15 represent part of a nut provided with a United States standard thread. The screw—not shown in the illustration—is supporting a load W, which is resolved into the two reactions R, one for each half of the nut. As the forces R are normal to any single element of the thread surface, the angle $\frac{u}{2}$ between R and W will be $60° \div \mathbf{2} = 30°$. The reaction R may again be resolved into two forces, one $\frac{W}{2}$ parallel with the axis of the bolt and the other Q, perpendicular to the axis. Then, $\frac{W}{2}$ represents the portion of the load carried by the surface a of the screw thread, while Q is the force that tends to burst the nut. For a given load W the force $\frac{W}{2}$ will remain constant whatever the angle u of the thread may be. On the other hand, the forces R and Q will increase as the angle u increases, and as the friction between two surfaces is proportional to the normal pressure between them, it follows that *the greater the angle of a screw thread, the greater is the friction between the threads of the bolt and the nut, and also the greater is the force tending to burst the nut.*

The relation between the total load W and the bursting pressure Q may be found as follows:

$$\frac{W}{2} = R\cos\frac{u}{2};\ W = 2R\cos\frac{u}{2};\ R = \frac{W}{2\cos\frac{u}{2}}$$

$$Q = R\sin\frac{u}{2} = \frac{W\sin\frac{u}{2}}{2\cos\frac{u}{2}} = \frac{W}{2}\tan\frac{u}{2} \qquad \textbf{(1)}$$

When $\frac{u}{2} = 30°$,

$$Q = \frac{W}{2}.577 = .289W$$

In this instance, the friction between the threads on the bolt and the nut has not been considered. This friction would tend to decrease the pressure Q, the reduction depending on the coefficient of friction between the contact surfaces. Assuming a coefficient of .132, the angle of friction e would be the angle of which .132 is the tangent, or 7° 30′. The angle $\frac{u}{2}$ would be reduced by this amount, making

$$Q = \frac{W}{2} \tan\left(\frac{u}{2} - e\right) = \frac{W}{2} \tan 22° \, 30' = .207\, W \qquad \textbf{(2)}$$

The pressure Q is shown in Fig. 15, as referring only to the one thread marked a, but it should be understood that the pressure is divided among all the threads and along the whole circumference of the nut. In calculating the resistance of the nut against the bursting pressure, the minimum cross-sectional area will be considered, which is that defined by a longitudinal section taken along the short diameter of the nut.

According to formula **2**, the pressure Q is equal to .207 of the total load supported by the bolt. The required sectional area of the nut should therefore be at least .207 of the bolt area for the coefficient of friction assumed. The equations that follow will show whether this condition is fulfilled by the nut.

Let d = diameter of bolt;
D = short diameter of nut;
h = height of nut = d.

Then, the cross-sectional area of the nut is $d(D - d)$. This area should be at least equal to .207 times the bolt area, or,

$$d(D - d) = .7854\, d^2 \times .207$$

Solving for D, $D - d = .1626\, d$

and $D = 1.1626\, d \qquad \textbf{(3)}$

It will be shown in a subsequent article that the diameter D for standard nuts is $D = 1\frac{1}{2}\, d + \frac{1}{16}$. This is in excess of that required by formula **3**, and the sectional area of the nut is therefore well within the safe limit. In the foregoing considerations it was assumed that the full diameter of the bolt was its effective diameter and that the height of the nut was equal to d.

32. In the square thread, the angle between the sides is zero; hence, there is no force tending to burst the nut. The reaction R at one-half of the nut becomes equal to $\frac{W}{2}$; therefore, the friction of a square thread is less than that of a triangular thread. On the other hand, the triangular thread is nearly twice as strong as a square thread. This will be seen more clearly by referring to Fig. 16, which illustrates a section of two bolts *a* and *b* of the same diameter, *a* having a square and *b* a United States standard thread. Since the number of threads per inch of the United States standard screw is twice as great as that of the square-threaded screw, the pitch p of the former is half that of the latter. The shearing surface of a single triangular thread is πd_1 multiplied by the distance $l n$, or $\pi d_1 p$, nearly, while the shearing surface of a single square thread is $\pi d_1 \frac{p}{2} = \frac{\pi d_1 p}{2}$

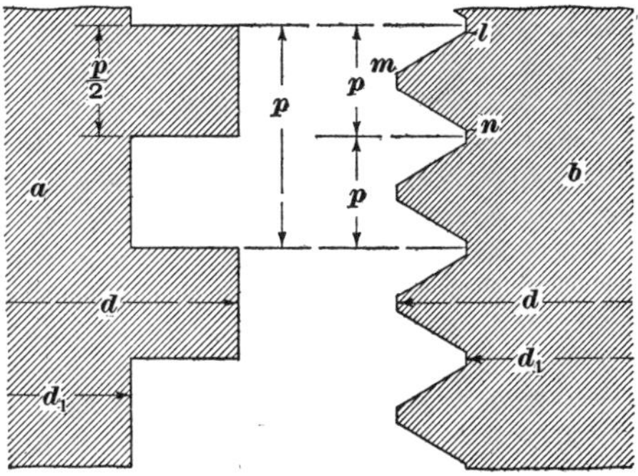

Fig. 16

It follows, therefore, that the triangular thread is better for fastenings, and the square thread, for transmitting motion.

33. The trapezoidal thread combines the good features of both the triangular and the square threads, having the same shearing section as the former and the same friction as the latter. Care should be taken, however, to use the screw so that the pressure will come on the flat side of the thread, for if it is put on the inclined side, the friction and bursting force on the nut will both be greater than for a 60° triangular thread.

34. Multiple-Threaded Screws.—A screw intended to transmit motion will sometimes have a very large pitch. As the depth of the thread in a square-threaded screw is equal to one-half the pitch, it is obvious that in some cases this depth will exceed the radius of the screw. This difficulty may be overcome by reducing the dimensions of the thread and by adding one or more threads in the manner about to be explained.

Assume that Fig. 17 is a single-threaded screw having an outside diameter d, a root diameter d_1, and a pitch p. It is desired to increase the diameter d_1 to d_2 and at the same time retain the diameter d, as shown in Fig. 18. A decrease of the depth of thread from t to t_1 is necessarily followed by a decrease in its width. Hence, the original cross-section of the thread, as indicated by the square $a\,b\,d\,c$, shown dotted, is reduced to the square $a\,b_1\,d_1\,c_1$. The distance between similarly located points on the thread represents, as before, the pitch p.

When it is simply a matter of producing a desired motion, a screw with the amount of space between the threads shown in the left half of the figure would not be objectionable. But

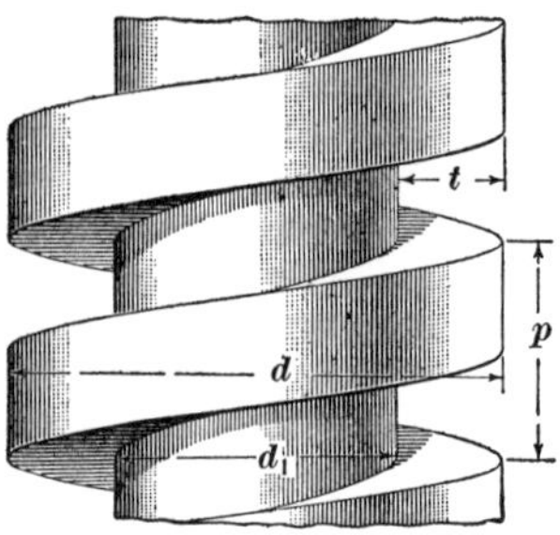

Fig. 17

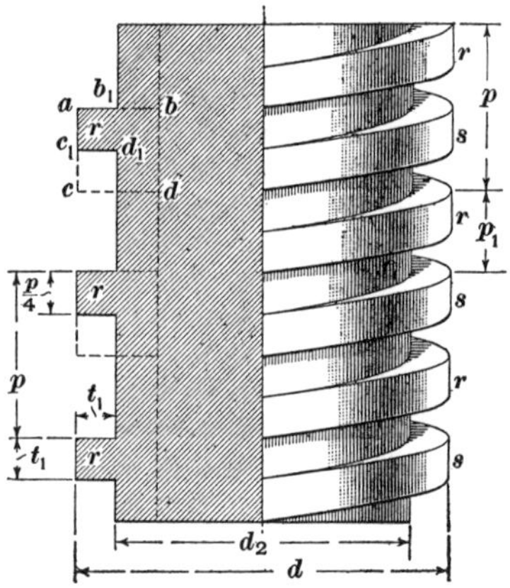

Fig. 18

when a load is to be supported, it is desirable to increase the shearing resistance of the thread as well as its bearing surface. This may be accomplished by adding another thread, as shown in the right half of the figure, where r represents the first thread and s the supplementary one.

A further increase in the diameter d_2 would make it possible to add more threads. A screw with more than one thread is termed a **multiple-threaded screw;** one with two threads, a **double-threaded screw;** with three threads, a **triple-threaded screw;** and with four threads, a **quadruple-threaded screw.**

The object of using a multiple-threaded screw, therefore, is to obtain a certain desired motion for each revolution of the screw, and at the same time to have ample strength to move the load. A single-threaded screw of the same lead would seldom have the required strength.

35. The number of threads n on a multiple-threaded screw is inversely proportional to the reduction in width of the thread of a single-threaded screw having the same pitch. For instance, if the thread of a single-threaded screw has a width of $\frac{1}{2}p$, and it is desired to increase the number of threads to n, then the width of each thread in the multiple-threaded screw will be $\frac{p}{2n}$.

In the case of multiple-threaded screws, the *lead* and the *pitch* are not equal. Instead, the pitch is equal to the lead divided by the number of threads. Thus, if the lead of a quadruple-threaded screw is 6 inches, the pitch is $6 \div 4 = 1\frac{1}{2}$ inches. In connection with multiple-threaded screws, the term *divided pitch* is often used in place of pitch. Thus, if the lead of a triple-threaded screw is $4\frac{1}{2}$ inches, the divided pitch is $4\frac{1}{2} \div 3 = 1\frac{1}{2}$ inches. In Fig. 18, the distance p is the lead and p_1 is the pitch, or, as it is sometimes called, the divided pitch. The dimensions of the thread are based on the divided pitch; that is, $t_1 = \frac{1}{2}p_1$.

36. By increasing the root diameter of a multiple-threaded screw, and consequently the number of threads, both the shearing and the helical surfaces of the threads are increased. Hence, the screw threads have a greater shearing strength, and the wear is divided over a greater area.

37. Pipe Threads.—The main difference between the pipe thread, known also as the **Briggs standard,** and the United States standard thread is that the former is built up on a conical instead of a cylindrical body. Besides being tapered, the thread must also be finer than the United States standard thread, as the depth of the latter in most cases would exceed the thickness of the pipe. For instance, the thickness of the walls in a 6-inch pipe is .280 inch, while the depth of a 6-inch screw thread, according to the United States standard, is .289 inch, thus exceeding the thickness of the pipe by .009 inch.

In a screw thread made according to the Briggs standard, the threads have an angle of 60° and are rounded off slightly at top and bottom, so that the depth of the thread is .80 of the pitch of the thread. After cutting the thread, the outside surface has a taper to a certain distance from the end, the standard taper being such that the inclination of the outside surface toward the axis of the pipe is 1 in 32. This makes the total taper, as measured by the variation in the outside diameter, equal to 1 in 16, or $\frac{3}{4}$ inch to the foot. The total length l, Fig. 19, of the tapered part is given by the empirical formula

$$l = \frac{4.8 \text{ inches} + .8\,d}{n}$$

in which d is the actual external diameter of the tube in the straight part, in inches, and n the number of threads per inch.

Let t_1 represent the depth of the rounded thread and p the pitch, then

$$t_1 = .8\,p = \frac{.8}{n}$$

The tapered part $a\,b$ of the pipe has threads that are perfect at top and bottom. At the beginning of the straight part, from b to $b\,a_1$, there are two threads that have the correct taper at the bottom, but their tops are flattened. The remainder of the threaded part $b_1\,b_2$ consists of four threads that are imperfect at both top and bottom. The latter are not essential to the Briggs system, but are simply a result of the method of cutting the thread at a single operation.

Table V, at the end of this Section, gives the dimensions of standard pipe threads. By means of the data given under Total Length of Thread and Length of Perfect Thread, it is possible to determine the length of the part containing imperfect threads. This length can also be determined directly, when it is remembered that the number of imperfect threads are always six, and that the space occupied by them depends on the number of threads per inch for that particular diameter.

Fig. 19 shows a longitudinal section through the threaded end of a 4-inch pipe, which, as seen from Table V, has an actual outside diameter of 4.5 inches. The taper of the threaded surface does not extend beyond the point b, at which place the pipe resumes its full diameter. The inclination of the outside surface of the pipe toward its axis in this instance brings the point a at a distance of .0328 inch below the original surface. Between the points b and b_1 are located the two threads that are imperfect at the top, and between b_1 and b_2 are those that are imperfect at both top and bottom. It will be noticed that all of the perfect thread bottoms lie along a line $a_1 c$, parallel with a line $a b$ drawn through the tops of the threads. The imperfect bottoms lie along a line that passes from a_2 to b_2.

38. The cause of this peculiar thread form will be understood by an examination of Fig. 19. The pipe begins to

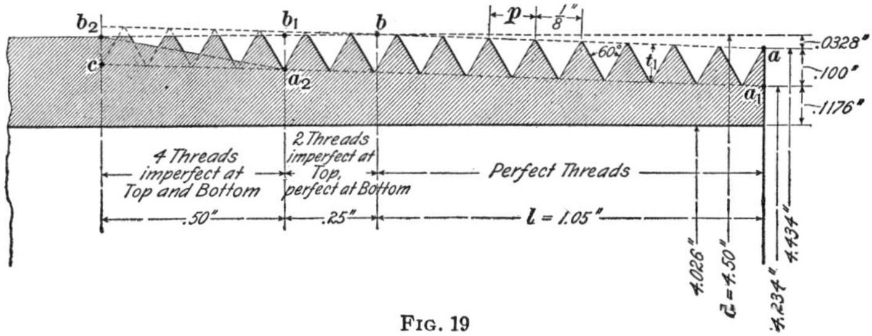

FIG. 19

assume a cylindrical form at point b, and since the die forming the threads is conical, the tops of the threads from b to b_2 cannot be filled out. The reason that the bottoms of threads are incomplete from points a_2 to c is due to the fact that parts of the die threads at this place are ground away in order that the die may be easily started.

From the method of cutting the threads, it follows that if the pipe is not passed far enough into the die, the number of perfect threads will be below the requirements. On the other hand, if the pipe is passed in too far, the extreme end of the pipe will be straight instead of tapering. In either case, it will be impossible to make a strong and tight joint.

BOLTS AND NUTS

STRENGTH OF SCREW BOLTS

39. Usually the stress on a bolt acts in the direction of its axis; that is, the bolt is in tension.

Let W = load on bolt, in pounds;
S_1 = *safe* working stress, in pounds per square inch;
a = area of cross-section of bolt at root of thread;
d = nominal (outside) diameter of bolt, in inches;
d_1 = diameter at root of thread, in inches.

Then, if the bolt is in tension,

$$W = a\,S_1;\ \text{or, } a = \frac{W}{S_1}$$

After finding the value of a, the corresponding values of the nominal diameter d and the diameter d_1 at bottom, or root, of the United States standard thread may be found from Table II. The relation between d and d_1 is also given in formula **4,** Art. **22.**

For bolts subjected to a constant tension, S_1 may be 8,000 pounds per square inch. More often the tension varies between zero and its maximum value; in this case, S_1 may be taken as 6,000 pounds per square inch. For cylinder-head bolts, and, in general, for bolts used to make a steam-tight joint, S_1 may vary from 3,000 pounds for small cylinders to 6,000 pounds for very large ones. Ordinarily, S_1 may be taken as 4,000 or 4,500 pounds per square inch. All the values just given are for wrought-iron bolts.

EXAMPLE.—Find the diameter of a wrought-iron bolt that is to sustain a steady load of $4\frac{1}{2}$ tons.

SOLUTION.—From the formula,

$$a = \frac{4\frac{1}{2} \times 2{,}000}{8{,}000} = 1.125 \text{ sq. in.}$$

From Table II, the value of d lies between $1\frac{3}{8}$ in. and $1\frac{1}{2}$ in. The latter value should be taken. Ans.

40. For screws that have square or Acme standard threads and that are intended to transmit motion, the following formula may be used, assuming that d_1 is approximately equal to $\frac{4}{5}d$:

$$W = 3{,}000\,d_1^2 = 1{,}920\,d^2$$

or

$$\left.\begin{aligned} d_1 &= .0183\sqrt{W} \\ d &= .0228\sqrt{W} \end{aligned}\right\} \qquad (1)$$

For screws of this character, the least number of threads in the nut that are necessary to prevent excessive wear is given by the following formula, in which n_1 is the number of threads in the nut:

$$n_1 = \frac{W}{300\,d_1^2} = .0052\frac{W}{d^2} \qquad (2)$$

Formula **2** applies to square and trapezoidal threads, and is based on the assumption that the pressure on the thread per square inch of projected area should not be greater than 700 pounds per square inch.

EXAMPLE.—A square-threaded screw $1\frac{1}{2}$ inches in diameter transmits motion to a load of 4,000 pounds. What is the least allowable number of threads in the nut?

SOLUTION.—By formula **2**,

$$n_1 = \frac{.0052 \times 4{,}000}{(1\frac{1}{2})^2} = 9\tfrac{1}{4}. \text{ Ans.}$$

PROPORTIONS OF BOLTS AND NUTS

41. The dimensions of the nut and bolt head depend on the diameter of the bolt. The United States standard form of bolt and nut, as shown in Fig. 20, has a square head and a hexagonal nut with washer, although the head may also be hexagonal. The washer is used to give a smooth seat for the nut to be screwed up against.

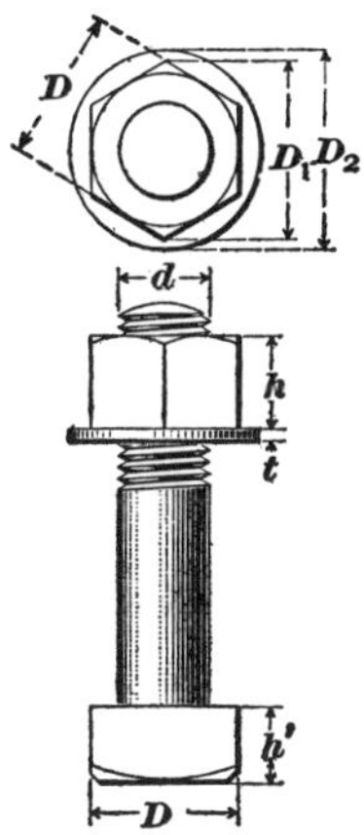

FIG. 20

The following proportions are usually adopted, the dimensions being identical for both rough and finished work. In other words, the parts that are to receive a finish are made larger in the rough, so that when finished they will correspond in size with those that are to remain rough.

The diameter of nut or head across flats is found by the formula

$$D = 1\tfrac{1}{2}d + \tfrac{1}{8} \text{ inch} \qquad (1)$$

This proportion holds for both hexagonal and square nuts. The diameter across corners D_1 may be found from the geometry of the figure. Thus, for hexagonal nuts,

$$D_1 = \frac{D}{\cos 30°} = \frac{D}{.866} = 1.73\,d + .14 \text{ inch} \qquad (2)$$

For square nuts,

$$D_1 = \frac{D}{\cos 45°} = D\sqrt{2} = 2.12\,d + .18 \text{ inch} \qquad (3)$$

For the height of nut,

$$h = d \qquad (4)$$

For the height of head,

$$h' = \frac{D}{2} = \tfrac{3}{4}d + \tfrac{1}{16} \text{ inch} \qquad (5)$$

The standard dimensions of plate washers are given in Table X at the end of this Section.

EXAMPLE.—Required, the various dimensions of a finished bolt and hexagonal nut, the bolt being $1\frac{1}{2}$ inches in diameter.

SOLUTION.—By the formulas just given, the diameter of the nut across flats is equal to

$$D = 1\tfrac{1}{2}d + \tfrac{1}{8} \text{ in.} = 1\tfrac{1}{2} \times 1\tfrac{1}{2} + \tfrac{1}{8} = 2\tfrac{3}{8} \text{ in. Ans.}$$

The diameter of the nut across corners is

$$D_1 = \frac{D}{.866} = \frac{2.375}{.866} = 2.742, \text{ or } 2\tfrac{3}{4}, \text{ in., nearly. Ans.}$$

The side of the square bolt head is $D = 2\frac{3}{8}$ in. Ans.

The height of the nut is $h = d = 1\frac{1}{2}$ in. Ans.

The height of the bolt head is

$$h' = \frac{D}{2} = \frac{2\frac{3}{8}}{2} = 1\tfrac{3}{16} \text{ in. Ans.}$$

From Table X, the diameter of the washer is $D_2 = 3\frac{1}{2}$ in., and the thickness t is $\frac{11}{64}$ in. Ans.

42. Table II, which will be found at the end of this Section, gives the dimensions of the United States standard bolts and nuts, and is based on the formulas given in Arts. **21**, **22**, and **41**. In regard to the number of threads per inch for diameters between those given in the table, it is customary to give them the number belonging to the next larger diameter. For instance, a bolt $4\frac{1}{16}$ inches in diameter will have $2\frac{7}{8}$ threads per inch.

The table, as far as it refers to threads, is generally complied with among the manufacturers, but many variations are found in the proportions of the heads and nuts. To a great extent, these parts are now made according to the *Manufacturers' standard.* In this system, the height h' of the rough hexagonal head is equal to d, and the height of the nut h, to $d + \frac{1}{8}$ inch. The dimensions of the nuts across the flats are larger than those given in Table II.

43. A. L. A. M. Standard Screws and Nuts.—The pitch of the United States standard thread has proved to be too coarse for screws used in light machinery. It has also been desirable in such cases to have the dimensions of the nuts and heads somewhat reduced. As the number and variety of these special threads and nuts have caused a great deal of inconvenience, the Association of Licensed Automobile Manufacturers has adopted a new standard for screws and nuts, the dimensions of which are given in Table VI. The form of thread used, however, is identical with that of the United States standard.

It is assumed that where the screws are to be used in soft material, such as cast iron, brass, bronze, or aluminum, the existing United States standard pitches will be used.

The following terms have been adopted by the Association: *Screw* is intended to supplant the present so-called coupling bolt and capscrew; *plain hexagonal nut* is to be used in place of the present so-called United States standard nut; *castle nut* is a name given to a new nut intended to be used where a positive locking system is desired and *facing* refers to a relieved portion under the screw heads, castle nuts, and plain nuts.

The screw heads and plain nuts must be flat-chamfered, as shown in the diagrams at the top of Table VI, which also shows the form of locknut, known as a castle nut. Three diametral grooves in the latter permit a very close adjustment, and it is locked by means of a pin inserted through a hole in the bolt. The castle nut is also chamfered and casehardened. The screws, screw heads, and plain nuts are left soft, and all heads and nuts are to be semifinished.

WRENCHES

44. The usual forms of solid wrenches are illustrated in Fig. 21, that shown at (*a*) being used for hexagonal nuts, and that at (*b*) for square nuts. The length may be from fifteen to eighteen times the diameter of the bolt for which

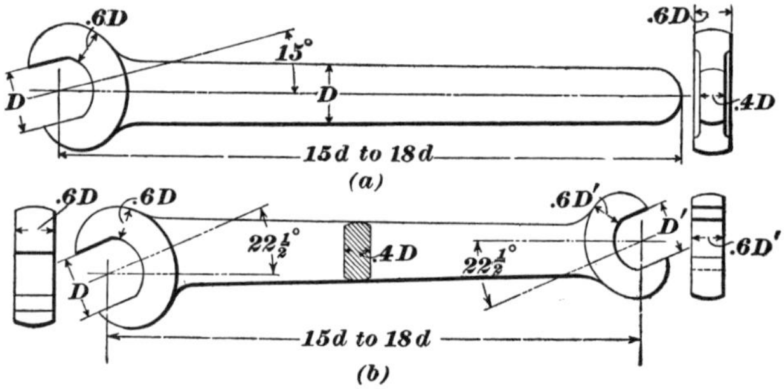

FIG. 21

it is to be used. In the figures, *d* represents the diameter of the bolt. The other proportions are given in terms of the diameter across the flats of the nuts, as shown in the figure.

The advantage derived from placing the opening of a wrench at an angle with the handle is that the nut may be turned completely around in locations where the swing of the handle is limited. Thus, the wrenches (*a*) and (*b*) may turn their respective nuts a complete revolution, when the swings of the handles are limited to 30° and 45°, respectively. By turning the wrench upside down after each partial revolution, it is possible to place the handle in its initial position and thus gain room for as great an angular swing as the surrounding parts will allow.

FORMS OF BOLT HEADS

45. Common Forms.—The ordinary *square bolt head* has already been shown in Fig. 20. Other forms are shown in Figs. 22 to 30.

In Fig. 22, the **hexagonal bolt head** is similar to a hexagonal nut, and has the same dimensions, except that the height h' may be less. Usually, $h' = \frac{2}{3}d$ to d. Fig. 23 shows a hexagonal head with a collar or flange, which is added to give an increased bearing surface.

A **cylindrical head** is shown in Fig. 24, and a **hemispherical head** in Fig. 25. The height h' of the former may be from $.5\,d$ to $.8\,d$, while that of the latter is $\frac{3}{4}d$. The diameter of these heads is shown in the figures. Fig. 26 shows a bolt head with a hemispherical bearing surface

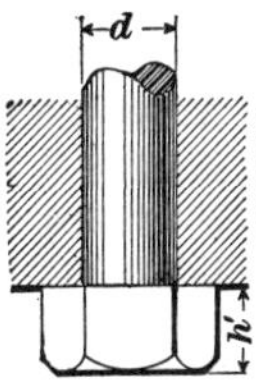

FIG. 22

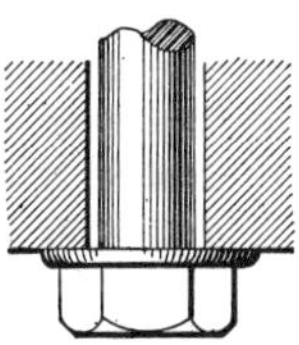

FIG. 23

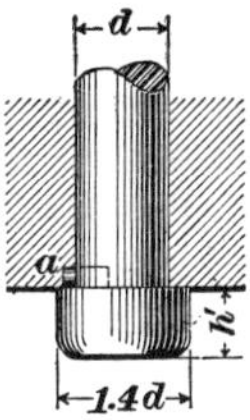

FIG. 24

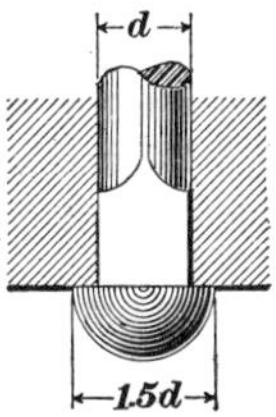

FIG. 25

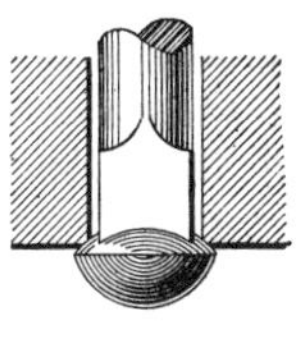

FIG. 26

resting on a seat of the same shape. This bolt may lean to one side, and the head will still remain in contact with its seat all the way around.

An **eyebolt** is shown in Fig. 27. The cross-section of the eye through the hole should either equal or exceed the area of the bolt; that is, referring to this figure,

$$2\,a\,b = .7854\,d^2;\text{ or, } a\,b = .39\,d^2$$

In good practice, $a\,b$ is equal to at least $\frac{1}{2}d^2$. On calculating the diameter d_1 of the pin passing through the eye, it should be observed that the pin is in double shear, the shearing surface being twice the area of it; or,

$$2\,(.7854\,d_1^2) = 1.57\,d_1^2$$

The strength of this pin in shear should equal the strength of the eyebolt in tension; therefore, letting S_s represent the *ultimate* shearing stress per square inch, S_t the *ultimate* tensile strength, and f the factor of safety,

$$1.57\, d_1^2 \frac{S_s}{f} = .7854\, d^2 \frac{S_t}{f}$$

or

$$d_1 = d\sqrt{\frac{S_t}{2\, S_s}} \qquad (1)$$

If, however, the pin is overhung, that is, if there is only one nib instead of the two shown at S, S, Fig. 27, it will be in single shear, and

$$d_1 = d\sqrt{\frac{S_t}{S_s}} \qquad (2)$$

Fig. 28 shows the head of a **hook bolt.** This form of bolt is used when it is undesirable to weaken one of the connected pieces by a bolt hole. The proportions are shown in the figure.

The **countersunk head** is shown in Fig. 29, and the **T head** in Fig. 30.

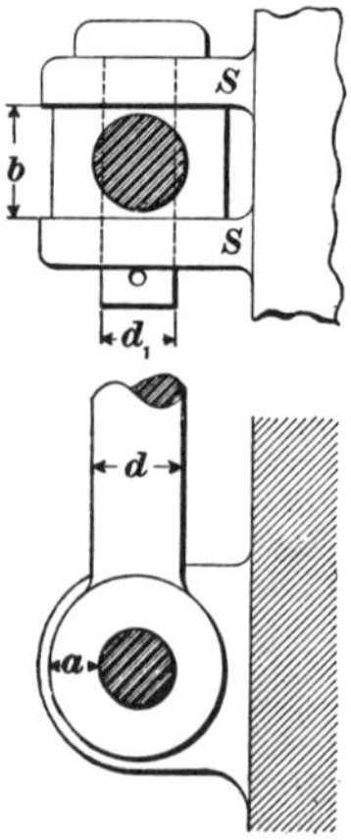

FIG. 27

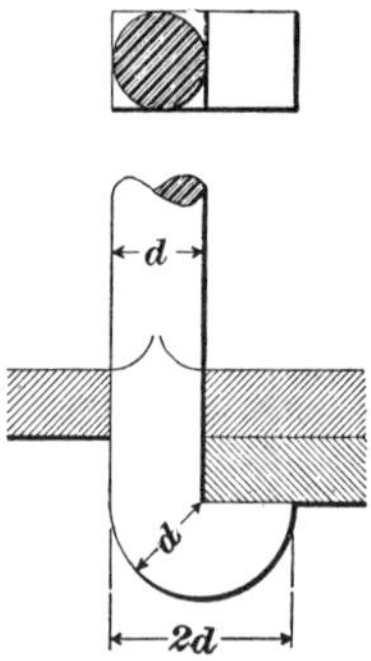

FIG. 28

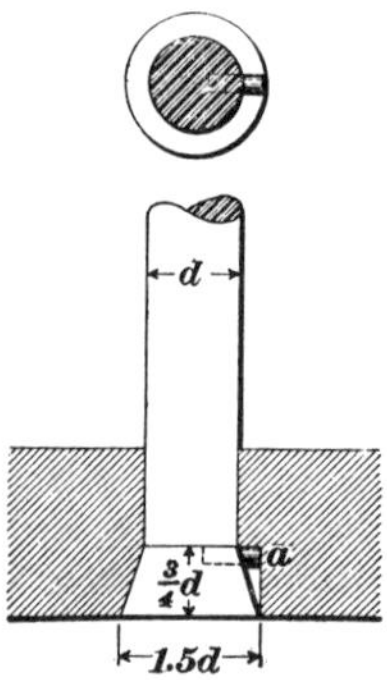

FIG. 29

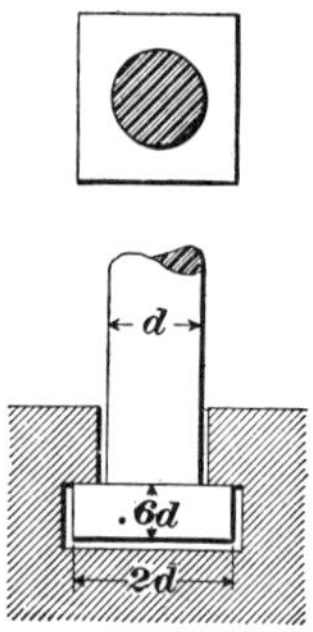

FIG. 30

46. Foundation Bolts.—The ordinary method of *attaching a bolt to stonework* is illustrated in Fig. 31. The head of the bolt is long and rectangular, and is made jagged with a cold chisel, while the hole in the stone is cut larger at the bottom than at the mouth. The bolt head is placed in the hole, and the remaining space is then filled with melted led or sulphur.

47. One method of securing the foundation bolts that fasten an engine bed to its foundation is shown in Fig. 32. These foundation bolts have no solid heads, but consist of long rods that are threaded on one end for a nut and have a slot in the other through which a *cotter* C passes. The cotter rests against a cast- or wrought-iron washer m, and the washer and cotter thus form the head of the bolt. The bolt head is within a recess formed in the foundation, which is arranged so as to be accessible.

The area of the washer bearing against the foundation, multiplied by the safe compressive strength of the material

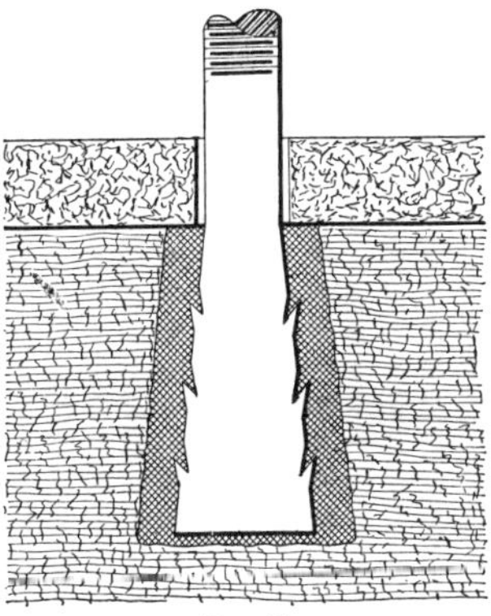

Fig. 31

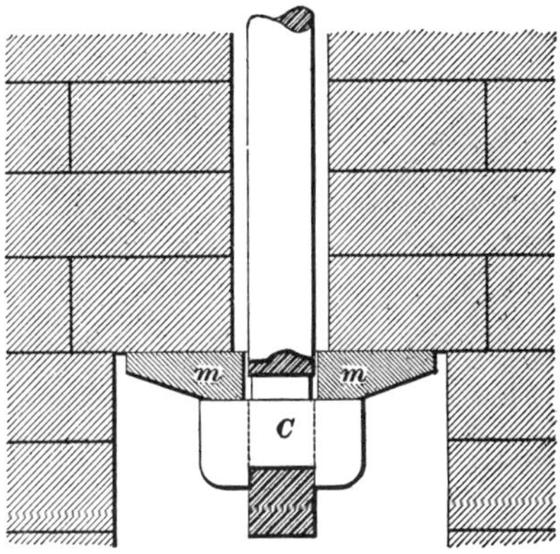

Fig. 32

of the foundation, should be equal to the tension carried by the bolt. For example, the tensile strength of wrought iron is about twenty times the compressive strength of brick. Hence, the bearing area of a washer resting against a brick foundation should be about twenty times the cross-section of the bolt.

48. Methods of Preventing Bolts From Turning. Various methods are used to prevent a bolt from turning while the nut is being screwed up. A common method is to make the neck of the bolt next to the head *square*, as shown in Figs. 25, 26, and 28. The bolt hole is also made square. Another way is to insert a pin *a* into the neck close to the head, as shown in Figs. 24 and 29. The projecting part of the pin fits into a recess cut out to receive it.

FORMS OF NUTS

49. Ordinary Forms.—The *common hexagonal nut* has been shown in Fig. 20. Ordinarily, both hexagonal and square bolt heads and nuts are chamfered off at an angle of 30° or 40°. Other forms of nuts are shown in Figs. 33 to 37.

The **flanged nut,** Fig. 33, is useful when the bolt hole is larger than the bolt, as it covers the hole and gives a greater

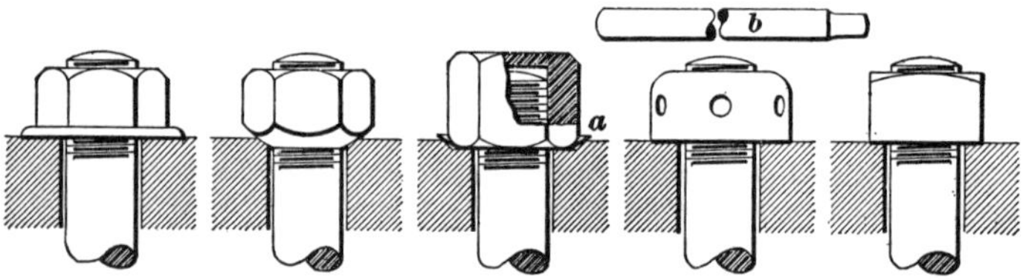

FIG. 33 FIG. 34 FIG. 35 FIG. 36 FIG. 37

bearing surface. While the same results may be obtained by means of a nut and washer, there are cases where it is preferable for the nut and washer to be in one piece.

Fig. 34 shows a nut with a **spherical bearing surface** and the seat shaped to correspond. This nut will bear on the seat all around, whether the bolt is perpendicular or inclined to the seat.

A **cap nut** is shown in Fig. 35. This form of nut is used to prevent fluids from leaking past the screw threads. To prevent leakage past the seat, the nut is screwed down on a soft, thin copper washer *a*.

Fig. 36 shows a **round nut,** and Fig. 37 an ordinary **square nut.** The round nut is provided with holes in its circumference, as shown, and is screwed up by inserting a bar *b* in one of the holes.

50. Round Slotted Nuts.—**Round slotted nuts** are used for bolts having large diameter or for shafts. They are provided with grooves as shown in Fig. 38, and are turned by means of a spanner engaging with these grooves. If D is the short diameter of a hexagonal nut, then the diameter of a round nut will be

$$D_c = D + 2\,s_2$$

s_2 being the depth of a groove. The height of the nut is $h = d$, while the width s_1 varies from 2 to 2.6 s_2. For a 1-inch bolt, $D_c = 2$ inches, $s_1 = \frac{3}{8}$ inch, and $s_2 = \frac{3}{16}$ inch.

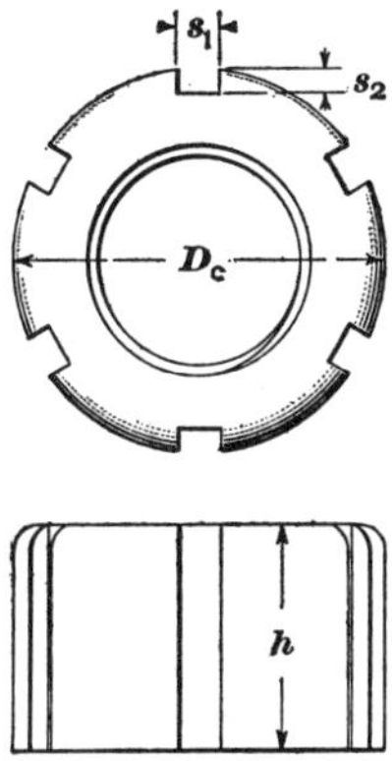

FIG. 38

LOCKING DEVICES

51. Locknuts.—All nuts are slightly loose on their bolts, a small clearance being necessary to permit them to turn freely. When a nut is subject to vibration it is likely to slack back and allow the bolt to become loose. To prevent this slacking back, various locking arrangements have been devised. A common device is the **locknut,** or **jam nut,** shown in Fig. 39. Two nuts are used, one of which is about half as thick as the ordinary nut. The load is thrown

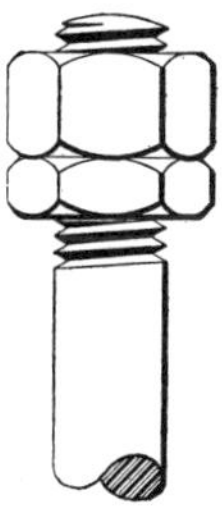

FIG. 39

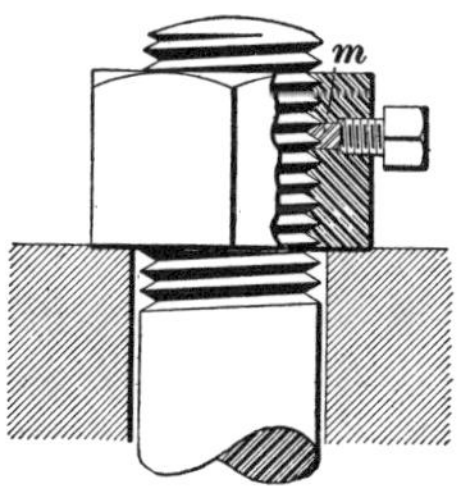

FIG. 40

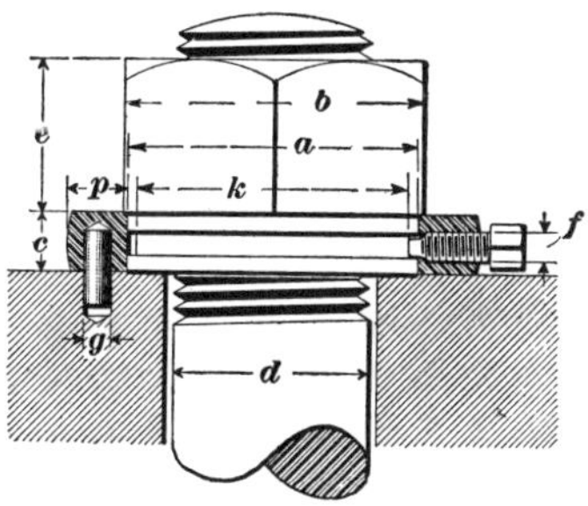

FIG. 41

on the outer nut, which should therefore be the thicker one. In practice, however, the thin nut is usually placed on the outside, because the wrench is generally too thick to act on it when placed below the other. The jam nut is not always satisfactory as a locking device.

52. Locking by Means of Setscrews.—The nut may be effectively locked to the bolt by the use of a setscrew, as shown in Fig. 40. To prevent the point of the setscrew from injuring the thread, a piece of iron or steel m may be let into the nut. This piece is screwed along with the nut, and acts as a shield interposed between the setscrew and the thread.

53. A good method of locking a nut is shown in Fig. 41. The lower portion of the nut is turned down, and a groove is cut in the center of the circular portion. A collar is fastened by means of a pin to one of the pieces to be connected, and the circular part of the nut is fitted into this collar. The nut is then bound to the collar by a setscrew passing through the latter, the point of the setscrew engaging into the groove turned in the nut. The following proportions, in which the diameter of the bolt d is taken as a unit, have proved very satisfactory:

$$a = 1\tfrac{1}{2}d - \tfrac{1}{16} \text{ in.}; \qquad f = \tfrac{1}{8}d + \tfrac{1}{8} \text{ in.};$$
$$b = 1\tfrac{1}{2}d + \tfrac{1}{8} \text{ in.}; \qquad g = \tfrac{1}{8}d + \tfrac{1}{16} \text{ in.};$$
$$c = \tfrac{1}{4}d + \tfrac{1}{4} \text{ in.}; \qquad p = \tfrac{1}{4}d + \tfrac{1}{4} \text{ in.};$$
$$e = \tfrac{3}{4}d; \qquad k = a - \tfrac{1}{8} \text{ in.}$$

A variation of this nut-locking device is shown in Fig. 42. By this method the collar is omitted, and the setscrew is inserted in one of the pieces to be connected. Both of these devices are very effective and are used for heavy nuts on quickly moving parts and on important bearings, as in marine engines.

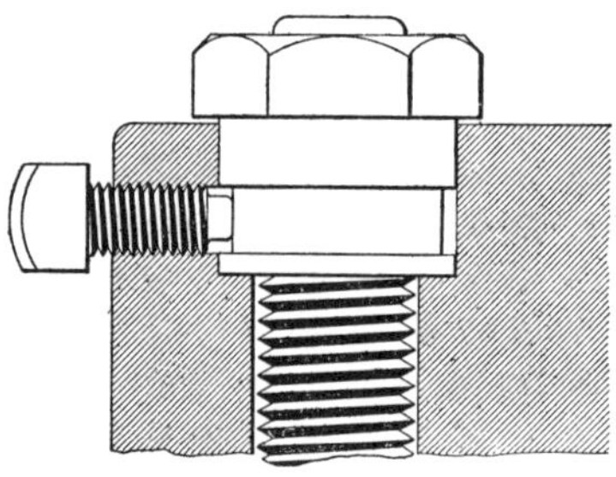

FIG. 42

54. Locking by Means of Stop-Plates.—Fig. 43 shows a device for locking a nut by means of a **stop-plate.** The plate is fastened to one of the pieces through which the bolt passes, and is so shaped that the bolt may be locked at intervals of $\frac{1}{12}$ revolution. Suitable proportions for this stop-plate are shown in the figure, in which d, the diameter of the bolt, is taken as the unit, with the exception of the distance between the center of the bolt and the center of the screw, for which D, the diameter of the nut across flats, is taken as the unit. All dimensions are in inches.

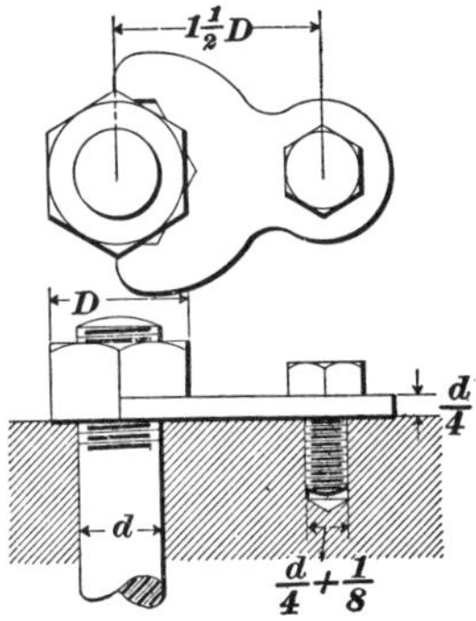

FIG. 43

55. In Fig. 44 is shown another form of stop-plate, which may be conveniently used when the bolts are set in a circle, as, for example, on engine cylinder heads.

56. Slotted Locknut.—In Fig. 45 is shown a different manner of locking the nut. In this method the nut is sawed half way through, and the parts are connected by a small screw. When the nut is screwed home the small screw is tightened, thereby greatly increasing the friction between the bolt and the nut.

57. Lock Washers.—A convenient locking device, called the **positive lock washer,** is shown in Fig. 46. When not held down by the nut, the washer has the form shown in (*b*); when the nut is screwed down tightly, as in (*a*), the washer is flattened out and its body carries the load of

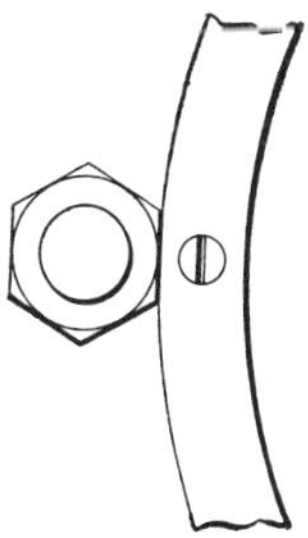

FIG. 44

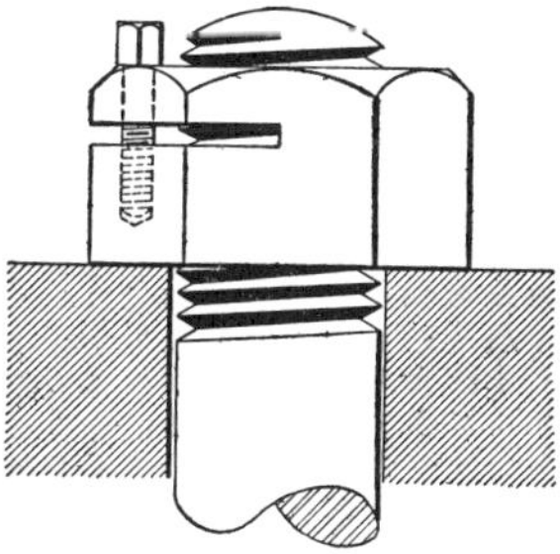

FIG. 45

compression. The tapered ends are thus relieved, and the spring tension is constant. As the barbs at the ends are free to move when subjected to vibration, they force themselves gradually into the positions indicated.

Other spring washers are the **Grover spring washer,** which has no barbs but simply depends on its elasticity to keep the nut tight on the bolt, and the **National lock washer,** which has, along its inner side next to the nut contact face, a rib that upsets and forces some of the metal of the nut into the bolt thread, and thus prevents the nut from loosening.

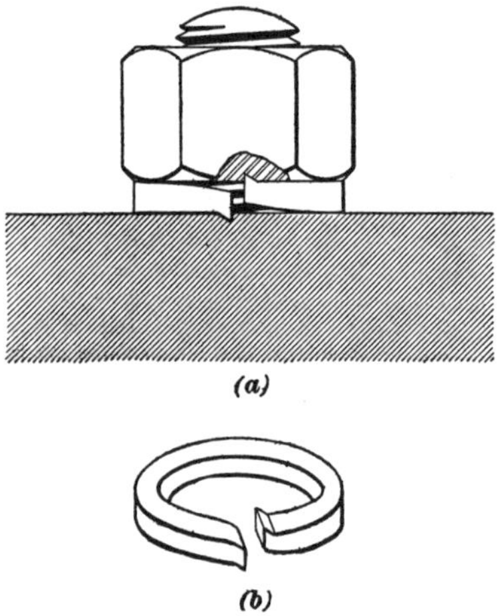

FIG. 46

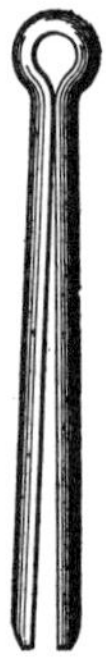
FIG. 47

58. Split Pins. **Split pins** may also be considered as locking devices. These pins are used where the parts occupy one position only, and from which they are seldom removed. After the nut is screwed into position, a hole is drilled through the *bolt* close to the nut, so that when the split pin is inserted it may prevent any motion of the nut. The split pin may be either straight or tapering; may be made of half-round wire and bent into the shape shown in Fig. 47; or may be circular in section and split at one end only. When in position, the projecting split ends are spread apart.

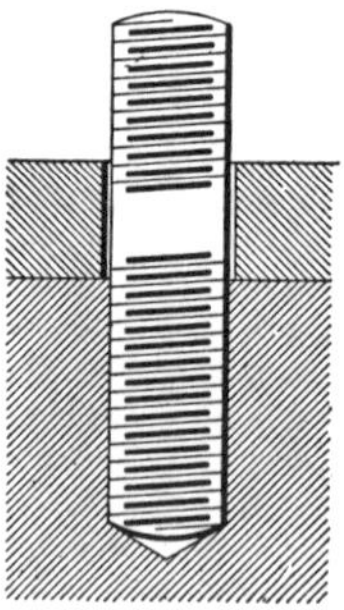
FIG. 48

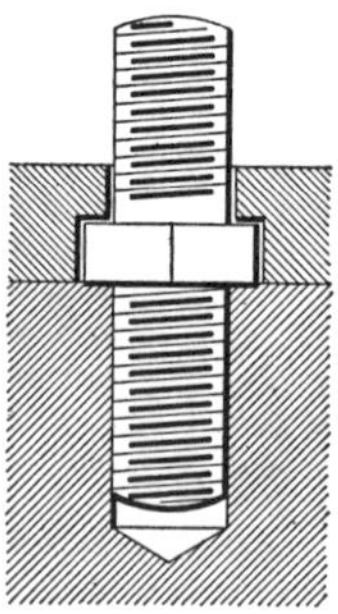
FIG. 49

FORMS OF BOLTS AND SCREWS

59. Bolts.—A **stud bolt, or stud,** is shown in Fig. 48. Each end of the stud has a screw thread cut on it. One end screws into one of the pieces to be connected, and the other carries a nut.

A stud having a *collar* is shown in Fig. 49. The collar, which may be either square or round, serves as a shoulder against which to screw up the stud. Also, when square, the collar is a convenient place to apply a wrench.

A **tap bolt,** shown in Fig. 50, is a bolt that does not require a nut. It is screwed directly into one of

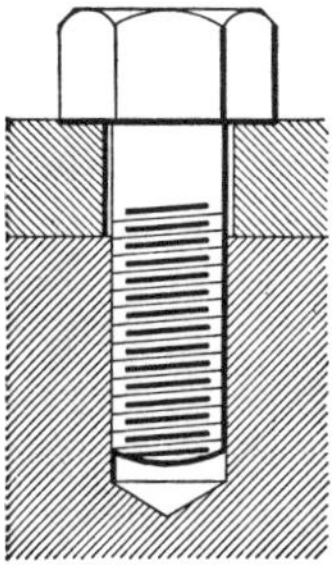

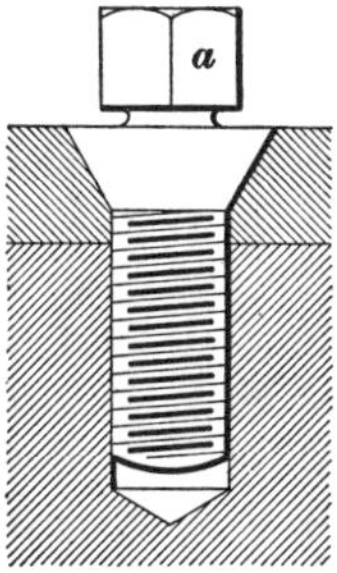

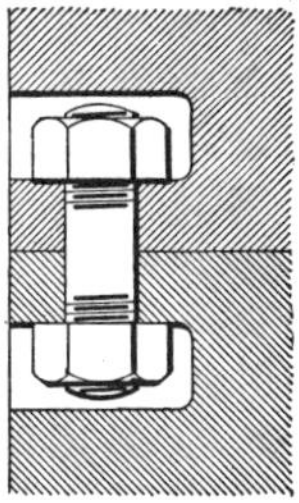

FIG. 50 FIG. 51 FIG. 52

the pieces to be connected, while the head presses on the other piece.

Fig. 51 shows a tap bolt having a countersunk head. This style of bolt is called a **patch bolt,** and, as its name implies, is used for making patches in boilers, etc. The diameter of the neck of the projection *a* to which the wrench is applied is smaller than the root diameter of the bolt, so that, instead of the bolt breaking when too much force is applied, the projection will break off.

Fig. 52 shows a bolt having a nut at each end instead of a head at one end and a nut at the other.

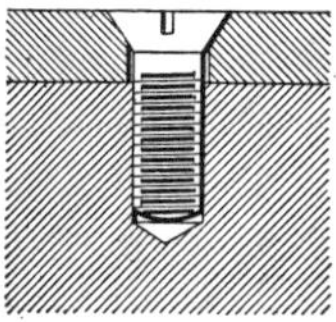

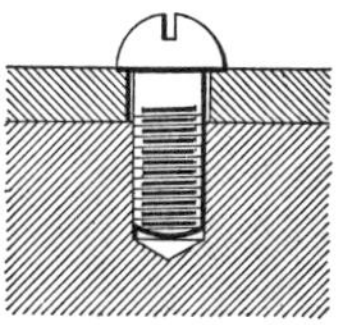

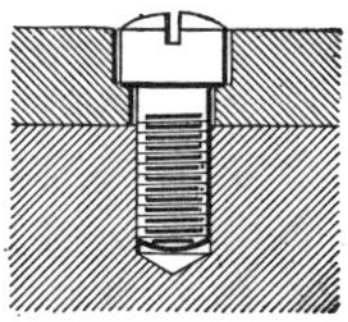

FIG. 53 FIG. 54 FIG. 55

60. Screws.—In Figs. 53, 54, and 55 are shown several forms of machine screws slotted for a screwdriver. The one shown in Fig. 53 is called a **countersunk-head screw,** or a **flathead machine screw;** that in Fig. 54, a **button-head screw;** and that in Fig. 55, a **fillister-head screw.** When a countersunk-head screw is used, the hole in the piece that is to be fastened is countersunk so that the head of the screw is flush, as shown.

61. Setscrews.—Screws or bolts that are used to press against a piece, and thus, by friction, prevent it from moving or rotating relatively to another piece are called **setscrews.** For example, a setscrew may be screwed through the hub of

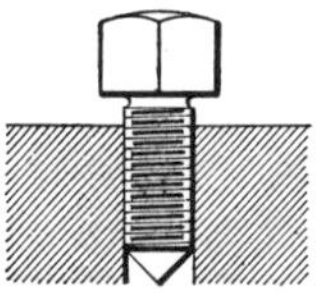
Fig. 56

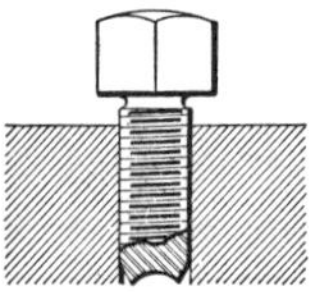
Fig. 57

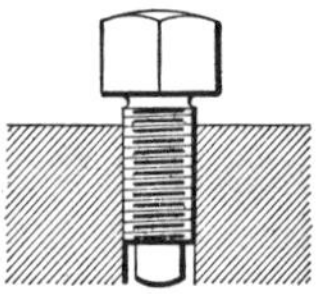
Fig. 58

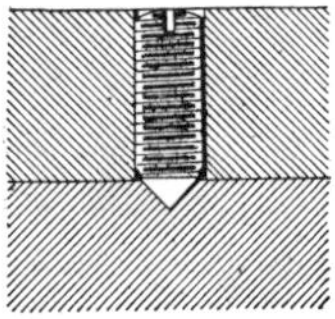
Fig. 59

a pulley, and by its pressure against the shaft will prevent the pulley from turning on the shaft.

Various forms of setscrews are shown in Figs. 56 to 59. The one shown in Fig. 56 is called a **cone-point setscrew;** in Fig. 57, a **cupped setscrew;** in Fig. 58, a **round pivot-point setscrew;** and in Fig. 59, a **headless cone-point setscrew.**

62. Bolts in Shear.—Usually bolts are in direct tension, but constructions occur in which a bolt may be placed in shear. The relative values of the tensile and shearing strengths of a material may vary. For wrought iron, these values may be equal, as indicated in *Strength of Materials.* But in order to be on the safe side, it may be supposed that in the material used for bolts the ultimate shearing strength is less than the tensile strength. Assuming that the shearing strength of wrought iron is about four-fifths of the tensile strength, the strength of a bolt in shear is about four-fifths that of a bolt in tension. Hence, the diameter of a bolt in single shear should be $\sqrt{\frac{5}{4}} = 1.1$ that of a bolt in tension under the same load, and the diameter of a bolt in double shear should be $\sqrt{\frac{5}{4 \times 2}} = .8$ that of a bolt in tension under the same load.

63. Knuckle Joint.—The **knuckle joint,** as shown in Fig. 60, is an example of a bolt in shear. Since the bolt is

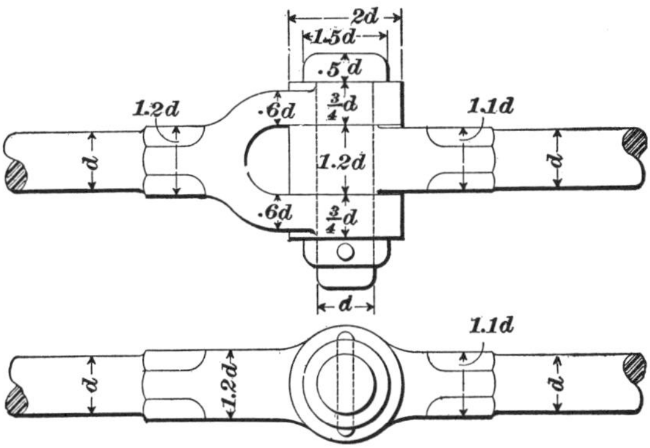

FIG. 60

in double shear, it need be theoretically only .8 the diameter of the rod. The bolt wears, however, and since it should at no time be less than .8 the diameter of the rod, the bolt and rod are made equal in diameter. The other proportions are in terms of the diameter of the bolt, all dimensions being in inches.

Other examples of bolts in shear may be seen in pin-connected iron bridges.

EXAMPLES FOR PRACTICE

1. Calculate the diameter of a wrought-iron bolt that is to sustain a varying load of 2,300 pounds. Ans. $\frac{7}{8}$ in.

2. What steady load may be safely sustained by five bolts $1\frac{1}{8}$ inches in diameter? Ans. 13.85 tons

3. A screw with square threads transmits motion to a load of 1,500 pounds. Calculate the diameter of the screw, the number of threads per inch, and the necessary number of threads in the nut.

Ans. { Diameter = $\frac{7}{8}$ in.
Threads per inch = $4\frac{1}{2}$
Threads in nut = 10

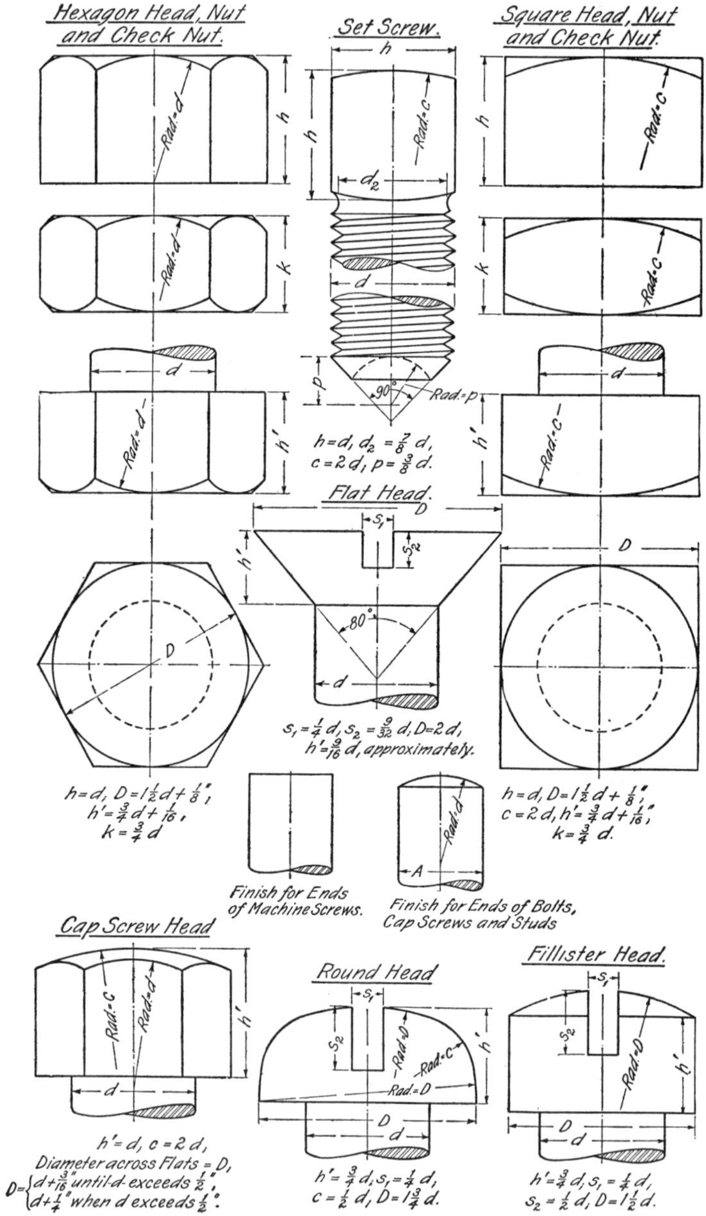
Hexagon Head, Nut and Check Nut.
$h=d,\ D=1\frac{1}{2}d+\frac{1}{8}''$,
$h'=\frac{3}{4}d+\frac{1}{16}$,
$k=\frac{3}{4}d$
Set Screw.
$h=d,\ d_2=\frac{7}{8}d$,
$c=2d;\ p=\frac{3}{8}d$.
Square Head, Nut and Check Nut.
$h=d,\ D=1\frac{1}{2}d+\frac{1}{8}''$,
$c=2d,\ h'=\frac{3}{4}d+\frac{1}{16}''$,
$k=\frac{3}{4}d$.
Flat Head.
$s_1=\frac{1}{4}d,\ s_2=\frac{9}{32}d;\ D=2d$,
$h'=\frac{9}{16}d$, approximately.
Finish for Ends of Machine Screws.
Finish for Ends of Bolts, Cap Screws and Studs
Cap Screw Head
$h'=d,\ c=2d$,
Diameter across Flats $=D$,
$D=\begin{cases} d+\frac{3}{16}'' \text{ until } d \text{ exceeds } \frac{1}{2}'' \\ d+\frac{1}{4}'' \text{ when } d \text{ exceeds } \frac{1}{2}'' \end{cases}$
Round Head
$h'=\frac{3}{4}d;\ s_1=\frac{1}{4}d$,
$c=\frac{1}{2}d,\ D=1\frac{3}{4}d$.
Fillister Head.
$h'=\frac{3}{4}d;\ s_1=\frac{1}{4}d$,
$s_2=\frac{1}{2}d;\ D=1\frac{1}{2}d$.

FIG. 61

64. Details of Construction.—The additional information required for the construction of the standard forms of bolt heads, nuts, and the principal screw heads is given in Fig. 61. The diagrams of the standard hexagon and square heads and nuts have been repeated from Table II, for the purpose of giving the radii of the curved outlines of some of the faces. The various dimensions are indicated by letters in the diagrams, and the relations between these dimensions are stated either below each separate diagram or below each series of diagrams. It will be seen that all dimensions are based on the diameters of the bolts or screws.

In order to avoid the calculations required for the use of the diagrams in regard to standard bolts and nuts, the values given in Table II may be used, the same letters in each case referring to the same dimensions. The dimensions of the parts not referred to in Table II, such as tap bolts, capscrews, and setscrews, are given in Table I. The diagrams at the top of the table contain the letters that represent the various dimensions inserted in the same. The values given in Table I are not according to any generally accepted standard, and the number of threads per inch will vary with different manufacturers.

65. Dimensions of small standard machine or wood screws are given in Table VII. The sizes of these screws are indicated by gauge numbers instead of by diameters, but for the purpose of comparison, the table also gives the diameter, in inches, corresponding to any of the gauge numbers. To draw the head of any one of these screws, find its diameter in the table and lay out the head according to the instructions given in Fig. 61.

66. Diameters of Small Holes.—The sizes of small holes in general are also indicated by gauge numbers instead of by their diameters. In this case, they receive their gauge number from the drill used for drilling them. Table VIII gives the gauge numbers and corresponding diameters of the Morse twist drills. This table will also serve to indicate the gauge numbers of ordinary steel wire according to the

Morse system, but, as will be observed, it differs from other wire gauges, such as the Stubs, Brown & Sharpe, etc.

67. Decimal Equivalents of Parts of 1 Inch. Table IX is intended to assist in finding the nearest 64th inch corresponding in value to a dimension given in thousandths of an inch.

The decimal fractions are printed in two sizes of type, the large size giving the exact value of the corresponding fractional part of an inch to the fourth decimal place. A given decimal fraction of an inch is rarely exactly equal to any of these values, but is either above or below, and the question then is to decide to which of two values it is the nearest, whether to the preceding or the succeeding one. For instance, it is desired to lay off the fraction .1330 inch in 64ths of an inch. The nearest decimal fractions are .1250 and .1406, and the question is which of these to choose. By means of the decimal fractions, printed in smaller type, this question may be answered at once, as the value of any one of these is the mean of the two adjacent fractions printed in larger type. If, therefore, any of the given decimal fractions is *above* the mean in value, it belongs to the succeeding decimal fraction; if *below*, to the preceding one. In this instance the mean fraction is .1328, and as .1330 is greater than this, .1406 inch, or $\frac{9}{64}$ inch will be chosen. In the same manner the nearest 64th inch corresponding to the decimal fractions .3670 inch and .8979 inch are found to be $\frac{23}{64}$ inch and $\frac{57}{64}$ inch, respectively.

TABLE I

DIMENSIONS OF TAP BOLTS AND SETSCREWS

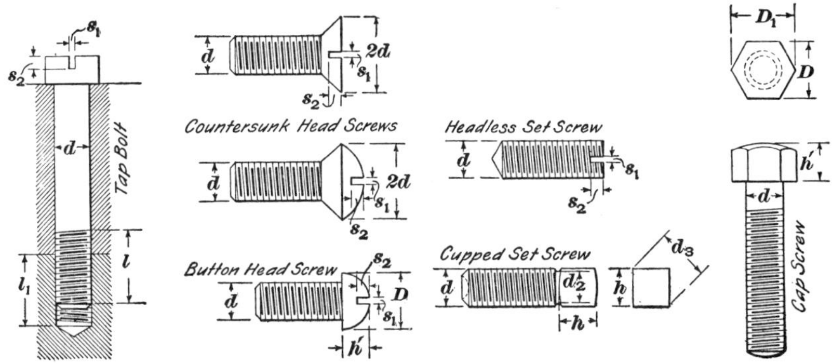

Diameter of Bolt	Number of Threads per Inch	Tap Bolts	Setscrews							Capscrews		
			Button Heads, Countersunk, and Headless				Cupped					
		Tap Hole l_1 Length of Thread	Diameter	Depth of Head	Width	Depth						
d	n	l_2	D	h'	s_1	s_2	h	d_2	d_3	D	D_1	h'
1/4	20	1/2	3/8	3/16	3/64	3/32	1/4	7/32	23/64			
5/16	18	5/8	15/32	15/64	1/16	1/8	5/16	9/32	7/16	1/2	37/64	5/16
3/8	16	3/4	9/16	9/32	1/16	1/8	3/8	21/64	17/32	9/16	21/32	3/8
7/16	14	7/8	5/8	21/64	3/32	3/16						
1/2	13	1	3/4	3/8	3/32	3/16	1/2	7/16	45/64	3/4	7/8	1/2
9/16	12	1 1/16	27/32	27/64	3/32	3/16						
5/8	11	1 3/16	15/16	15/32	3/32	3/16	5/8	35/64	7/8	7/8	1	5/8
3/4	10	1 3/8	1 1/16	9/16	1/8	1/4	3/4	21/32	1 1/16	1	1 5/32	3/4
7/8	9	1 9/16	1 1/4	21/32	1/8	1/4	7/8	49/64	1 15/64	1 1/8	1 5/16	7/8
1	8	1 3/4	1 7/16	3/4	5/32	5/16	1	7/8	1 13/32	1 1/4	1 7/16	1
1 1/8	7	1 15/16	1 5/8	27/32	5/32	5/16						
1 1/4	7	2 1/8	1 13/16	15/16	3/16	3/8				1 1/2	1 3/4	1 1/4
1 3/8	6	2 5/16	2	1 1/32	3/16	3/8						
1 1/2	6	2 1/2	2 3/16	1 1/8	7/32	7/16						
1 5/8	5 1/2	2 11/16										
1 3/4	5	2 7/8										
1 7/8	5	3 1/16										
2	4 1/2	3 1/4										
2 1/4	4 1/2	3 5/8										
2 1/2	4	4										
2 3/4	4	4 3/8										
3	4	4 3/4										

NOTE.—All dimensions are given in inches.

TABLE II

DIMENSIONS OF UNITED STATES STANDARD BOLTS AND NUTS

Bolts					Nuts					Heads
Diameter of Bolt	Number of Threads per Inch	Diameter at Bottom of Threads	Area at Bottom of Threads	Area of Body of Bolt	Diameter of Bolt, $d = h$ 	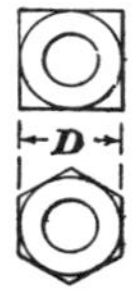	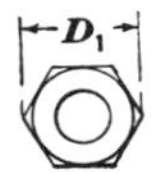	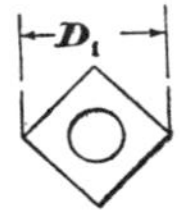	Diameter of Tap Drill	D and D_1 the Same as for Nuts 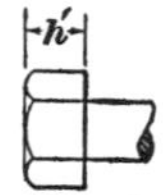
Inches d	n	Inches d_1	Square Inches a	Square Inches a_1	Rough Inches	Rough Inches	Rough Inches	Rough Inches	Inches	Rough Inches
$\frac{1}{4}$	20	.185	.0269	.0491	$\frac{1}{4}$	$\frac{1}{2}$	$\frac{37}{64}$	$\frac{45}{64}$	$\frac{3}{16}$	$\frac{1}{4}$
$\frac{5}{16}$	18	.240	.0452	.0767	$\frac{5}{16}$	$\frac{19}{32}$	$\frac{11}{16}$	$\frac{27}{32}$	$\frac{1}{4}$	$\frac{19}{64}$
$\frac{3}{8}$	16	.294	.0679	.1105	$\frac{3}{8}$	$\frac{11}{16}$	$\frac{51}{64}$	$\frac{31}{32}$	$\frac{5}{16}$	$\frac{11}{32}$
$\frac{7}{16}$	14	.345	.0935	.1503	$\frac{7}{16}$	$\frac{25}{32}$	$\frac{29}{32}$	$1\frac{7}{64}$	$\frac{23}{64}$	$\frac{25}{64}$
$\frac{1}{2}$	13	.400	.1257	.1964	$\frac{1}{2}$	$\frac{7}{8}$	$1\frac{1}{64}$	$1\frac{15}{64}$	$\frac{13}{32}$	$\frac{7}{16}$
$\frac{9}{16}$	12	.454	.1619	.2485	$\frac{9}{16}$	$\frac{9}{16}$	$1\frac{1}{8}$	$1\frac{3}{8}$	$\frac{15}{32}$	$\frac{31}{64}$
$\frac{5}{8}$	11	.507	.2019	.3068	$\frac{5}{8}$	$1\frac{1}{16}$	$1\frac{15}{64}$	$1\frac{1}{2}$	$\frac{17}{32}$	$\frac{17}{32}$
$\frac{3}{4}$	10	.620	.3019	.4418	$\frac{3}{4}$	$1\frac{1}{4}$	$1\frac{7}{16}$	$1\frac{3}{4}$	$\frac{5}{8}$	$\frac{5}{8}$
$\frac{7}{8}$	9	.731	.4197	.6013	$\frac{7}{8}$	$1\frac{7}{16}$	$1\frac{21}{32}$	$2\frac{1}{32}$	$\frac{3}{4}$	$\frac{23}{32}$
1	8	.838	.5515	.7854	1	$1\frac{5}{8}$	$1\frac{7}{8}$	$2\frac{19}{64}$	$\frac{27}{32}$	$\frac{13}{16}$
$1\frac{1}{8}$	7	.939	.6925	.9940	$1\frac{1}{8}$	$1\frac{13}{16}$	$1\frac{15}{16}$	$2\frac{9}{16}$	$\frac{31}{32}$	$\frac{29}{32}$

1 1/4	7	1.064	.8892	1.2272	1 1/4	2	2 5/16	2 53/64	1 3/32	1
1 3/8	6	1.158	1.0532	1.4849	1 3/8	2 3/16	2 17/32	3 3/32	1 3/16	1 3/32
1 1/2	6	1.283	1.2928	1.7671	1 1/2	2 3/8	2 3/4	3 23/64	1 9/32	1 3/16
1 5/8	5 1/2	1.389	1.5153	2.0739	1 5/8	2 9/16	2 61/64	3 5/8	1 13/32	1 9/32
1 3/4	5	1.490	1.7437	2.4053	1 3/4	2 3/4	3 11/64	3 57/64	1 1/2	1 3/8
1 7/8	5	1.615	2.0485	2.7612	1 7/8	2 15/16	3 25/64	4 5/32	1 5/8	1 15/32
2	4 1/2	1.711	2.2993	3.1416	2	3 1/8	3 39/64	4 27/64	1 3/4	1 9/16
2 1/4	4 1/2	1.961	3.0203	3.9761	2 1/4	3 1/2	4 3/64	4 61/64	1 31/32	1 3/4
2 1/2	4	2.175	3.7154	4.9087	2 1/2	3 7/8	4 15/16	5 31/64	2 3/16	1 15/16
2 3/4	4	2.425	4.6186	5.9396	2 3/4	4 1/4	4 29/32	6 1/64	2 7/16	2 1/8
3	3 1/2	2.629	5.4284	7.0686	3	4 5/8	5 11/32	6 35/64	2 5/8	2 5/16
3 1/4	3 1/2	2.879	6.5099	8.2958	3 1/4	5	5 25/32	7 5/64	2 29/32	2 1/2
3 1/2	3 1/4	3.100	7.5477	9.6211	3 1/2	5 3/8	6 13/64	7 19/32	3 1/32	2 11/16
3 3/4	3	3.317	8.6414	11.0447	3 3/4	5 3/4	6 41/64	8 1/8	3 11/32	2 7/8
4	3	3.567	9.9930	12.5664	4	6 1/8	7 5/64	8 21/32	3 19/32	3 1/16
4 1/4	2 7/8	3.798	11.3292	14.1863	4 1/4	6 1/2	7 1/2	9 3/16	3 13/16	3 1/4
4 1/2	2 3/4	4.027	12.7366	15.9044	4 1/2	6 7/8	7 15/16	9 23/32	4 1/32	3 7/16
4 3/4	2 5/8	4.255	14.2197	17.7206	4 3/4	7 1/4	8 3/8	10 1/4	4 9/32	3 5/8
5	2 1/2	4.480	15.7633	19.6350	5	7 5/8	8 13/16	10 25/32	4 1/2	3 13/16
5 1/4	2 1/2	4.730	17.5717	21.6476	5 1/4	8	9 15/64	11 5/16	4 3/4	4
5 1/2	2 3/8	4.953	19.2676	23.7584	5 1/2	8 3/8	9 43/64	11 27/32	4 31/32	4 3/16
5 3/4	2 3/8	5.203	21.2617	25.9673	5 3/4	8 3/4	10 7/64	12 3/8	5 7/32	4 3/8
6	2 1/4	5.423	23.0978	28.2744	6	9 1/8	10 17/32	12 29/32	5 7/16	4 9/16

TABLE III

INTERNATIONAL STANDARD SCREW THREAD

Bolt Diameter d		Diameter at Bottom of Thread d_1	Pitch p	Number of Threads per Inch	Bolt Diameter d		Diameter at Bottom of Thread d_1	Pitch p	Number of Threads per Inch
Mm.	Inches	Inches	Mm.		Mm.	Inches	Inches	Mm.	
6	.236	.181	1.00	25.4	33	1.299	1.106	3.50	7.26
7	.276	.220	1.00	25.4	36	1.417	1.196	4.00	6.35
8	.315	.246	1.25	20.3	39	1.535	1.314	4.00	6.35
9	.354	.285	1.25	20.3	42	1.654	1.404	4.50	5.64
10	.394	.311	1.50	16.9	45	1.772	1.523	4.50	5.64
11	.433	.350	1.50	16.9	48	1.890	1.613	5.00	5.08
12	.472	.376	1.75	14.5	52	2.047	1.770	5.00	5.08
14	.551	.441	2.00	12.7	56	2.205	1.900	5.50	4.62
16	.630	.519	2.00	12.7	60	2.362	2.058	5.50	4.62
18	.709	.571	2.50	10.16	64	2.520	2.187	6.00	4.23
20	.787	.649	2.50	10.16	68	2.677	2.345	6.00	4.23
22	.866	.728	2.50	10.16	72	2.835	2.474	6.50	3.91
24	.945	.779	3.00	8.47	76	2.992	2.632	6.50	3.91
27	1.063	.897	3.00	8.47	80	3.150	2.762	7.00	3.63
30	1.181	.987	3.50	7.26					

TABLE IV

ACME STANDARD SCREW THREAD

Threads per Inch n	Depth of Thread Inch t	Thickness at Top of Thread Inch a	Width of Space at Bottom of Thread Inch b	Width of Space at Top of Thread Inch c	Thickness at Root of Thread Inch e
1	.5100	.3707	.3655	.6293	.6345
$1\frac{1}{3}$	.3850	.2780	.2728	.4720	.4772
$1\frac{1}{2}$	.3433	.2471	.2419	.4195	.4247
2	.2600	.1853	.1801	.3147	.3199
3	.1767	.1235	.1183	.2098	.2150
4	.1350	.0927	.0875	.1573	.1625
5	.1100	.0741	.0689	.1259	.1311
6	.0933	.0618	.0566	.1049	.1101
7	.0814	.0529	.0478	.0899	.0951
8	.0725	.0463	.0411	.0787	.0839
9	.0655	.0413	.0361	.0699	.0751
10	.0600	.0371	.0319	.0629	.0681

TABLE V

UNITED STATES STANDARD STEAM, GAS, AND WATER PIPE

Sizes of Pipes Inches	Threads per Inch	Actual External Diameter Inches	Actual Internal Diameter Inches	Total Length of Thread Inches	Length of Perfect Thread Inches	Diameter of Tap Drill Inches	Diameter at End of Pipe Inches	
							Outside	At Bottom of Thread
$\frac{1}{8}$	27	.405	.270	.41	.19	$\frac{21}{64}$	.393	.334
$\frac{1}{4}$	18	.540	.364	.62	.29	$\frac{29}{64}$	.522	.433
$\frac{3}{8}$	18	.675	.494	.63	.30	$\frac{19}{32}$	.656	.568
$\frac{1}{2}$	14	.840	.623	.83	.39	$\frac{23}{32}$	.815	.701
$\frac{3}{4}$	14	1.050	.824	.84	.40	$\frac{15}{16}$	1.025	.911
1	$11\frac{1}{2}$	1.315	1.048	1.03	.51	$1\frac{3}{16}$	1.283	1.144
$1\frac{1}{4}$	$11\frac{1}{2}$	1.660	1.380	1.06	.54	$1\frac{15}{32}$	1.626	1.488
$1\frac{1}{2}$	$11\frac{1}{2}$	1.900	1.611	1.07	.55	$1\frac{23}{32}$	1.866	1.727
2	$11\frac{1}{2}$	2.375	2.067	1.10	.58	$2\frac{3}{16}$	2.339	2.200
$2\frac{1}{2}$	8	2.875	2.468	1.64	.89	$2\frac{11}{16}$	2.819	2.620
3	8	3.500	3.067	1.70	.95	$3\frac{5}{16}$	3.441	3.241
$3\frac{1}{2}$	8	4.000	3.548	1.75	1.00	$3\frac{13}{16}$	3.938	3.738
4	8	4.500	4.026	1.80	1.05	$4\frac{7}{32}$	4.434	4.234
$4\frac{1}{2}$	8	5.000	4.508	1.85	1.10	$4\frac{21}{32}$	4.931	4.731
5	8	5.563	5.045	1.91	1.16	$5\frac{15}{64}$	5.491	5.291
6	8	6.625	6.065	2.01	1.26	$6\frac{1}{4}$	6.546	6.346
7	8	7.625	7.023	2.11	1.36	$7\frac{5}{16}$	7.540	7.340
8	8	8.625	7.982	2.21	1.46	$8\frac{5}{16}$	8.534	8.334
9	8	9.625	8.937	2.32	1.57	$9\frac{5}{16}$	9.527	9.327
10	8	10.750	10.019	2.43	1.68	$10\frac{5}{16}$	10.645	10.445

NOTE.—The taper of the threaded part is $\frac{3}{4}$ inch per foot.

TABLE VI

ASSOCIATION OF LICENSED AUTOMOBILE MANUFACTURERS' STANDARD SCREWS AND NUTS

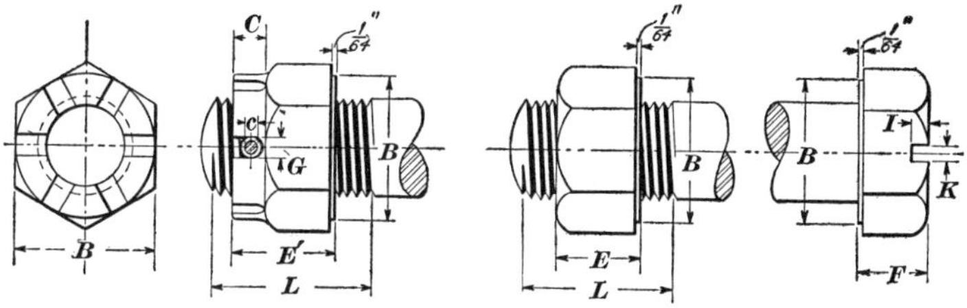

Diameter of Screw	Threads per Inch	Length of Threaded Portion	Short Diameter of Nut	Height of Castle	Thickness of Nut	Thickness of Castle Nut	Thickness of Head	Width of Slot for Cotter Pin	Depth of Slot in Head	Width of Slot in Head	Diameter of Cotter Pin	Diameter of Cotter-Pin Hole
d	*n*	*L*	*B*	*C*	*E*	*E'*	*F*	*G*	*I*	*K*	*c*	*G*
$\frac{1}{4}$	28	$\frac{3}{8}$	$\frac{3}{8}$	$\frac{3}{32}$	$\frac{7}{32}$	$\frac{9}{32}$	$\frac{3}{16}$	$\frac{5}{64}$	$\frac{3}{32}$	$\frac{1}{16}$	$\frac{1}{16}$	$\frac{5}{64}$
$\frac{5}{16}$	24	$\frac{15}{32}$	$\frac{1}{2}$	$\frac{3}{32}$	$\frac{17}{64}$	$\frac{21}{64}$	$\frac{15}{64}$	$\frac{5}{64}$	$\frac{7}{64}$	$\frac{1}{16}$	$\frac{1}{16}$	$\frac{5}{64}$
$\frac{3}{8}$	24	$\frac{9}{16}$	$\frac{9}{16}$	$\frac{1}{8}$	$\frac{21}{64}$	$\frac{13}{32}$	$\frac{9}{32}$	$\frac{1}{8}$	$\frac{1}{8}$	$\frac{3}{32}$	$\frac{3}{32}$	$\frac{1}{8}$
$\frac{7}{16}$	20	$\frac{21}{32}$	$\frac{11}{16}$	$\frac{1}{8}$	$\frac{3}{8}$	$\frac{29}{64}$	$\frac{21}{64}$	$\frac{1}{8}$	$\frac{1}{8}$	$\frac{3}{32}$	$\frac{3}{32}$	$\frac{1}{8}$
$\frac{1}{2}$	20	$\frac{3}{4}$	$\frac{3}{4}$	$\frac{3}{16}$	$\frac{7}{16}$	$\frac{9}{16}$	$\frac{3}{8}$	$\frac{1}{8}$	$\frac{1}{8}$	$\frac{3}{32}$	$\frac{3}{32}$	$\frac{1}{8}$
$\frac{9}{16}$	18	$\frac{27}{32}$	$\frac{7}{8}$	$\frac{3}{16}$	$\frac{31}{64}$	$\frac{39}{64}$	$\frac{27}{64}$	$\frac{5}{32}$	$\frac{1}{8}$	$\frac{3}{32}$	$\frac{1}{8}$	$\frac{5}{32}$
$\frac{5}{8}$	18	$\frac{15}{16}$	$\frac{15}{16}$	$\frac{1}{4}$	$\frac{35}{64}$	$\frac{23}{32}$	$\frac{15}{32}$	$\frac{5}{32}$	$\frac{1}{8}$	$\frac{3}{32}$	$\frac{1}{8}$	$\frac{5}{32}$
$\frac{11}{16}$	16	$1\frac{1}{32}$	1	$\frac{1}{4}$	$\frac{19}{32}$	$\frac{49}{64}$	$\frac{33}{64}$	$\frac{5}{32}$	$\frac{1}{8}$	$\frac{3}{32}$	$\frac{1}{8}$	$\frac{5}{32}$
$\frac{3}{4}$	16	$1\frac{1}{8}$	$1\frac{1}{8}$	$\frac{1}{4}$	$\frac{21}{32}$	$\frac{13}{16}$	$\frac{9}{16}$	$\frac{5}{32}$	$\frac{1}{8}$	$\frac{3}{32}$	$\frac{1}{8}$	$\frac{5}{32}$
$\frac{7}{8}$	14	$1\frac{5}{16}$	$1\frac{1}{4}$	$\frac{1}{4}$	$\frac{49}{64}$	$\frac{29}{32}$	$\frac{21}{32}$	$\frac{5}{32}$	$\frac{1}{8}$	$\frac{3}{32}$	$\frac{1}{8}$	$\frac{5}{32}$
1	14	$1\frac{1}{2}$	$1\frac{7}{16}$	$\frac{1}{4}$	$\frac{7}{8}$	1	$\frac{3}{4}$	$\frac{5}{32}$	$\frac{1}{8}$	$\frac{3}{32}$	$\frac{1}{8}$	$\frac{5}{32}$

NOTE.—The material is steel with an ultimate tensile strength not less than 100,000 pounds per square inch; the elastic limit is not less than 60,000 pounds per square inch; length of threaded portion is one and one-half times body diameter. All dimensions are given in inches.

TABLE VII

MACHINE AND WOOD SCREWS

Gauge No.	Diameter Inch	Threads per Inch	Diameter of Round Head Inch	Diameter of Filister Head Inch	Diameter of Flat Head Inch
000	.0315				
00	.0447				
0	.0578	104			
1	.0710	72			
2	.0842	64	.1544	.1332	.1631
3	.0973	48	.1786	.1545	.1894
4	.1105	36	.2028	.1747	.2158
5	.1236	32	.2270	.1985	.2421
6	.1368	32	.2510	.2175	.2684
7	.1500	32	.2754	.2392	.2947
8	.1631	32	.2936	.2610	.3210
9	.1763	32	.3238	.2805	.3474
10	.1894	32	.3480	.3035	.3737
11	.2026	24			
12	.2158	24	.3922	.3445	.4263
13	.2289	22			
14	.2421	20	.4364	.3885	.4790
15	.2552	20			
16	.2684	18	.4866	.4300	.5316
17	.2816	18			
18	.2947	18	.5248	.4710	.5842
19	.3079	18			
20	.3210	16	.5690	.5200	.6308
21	.3342				
22	.3474	16	.6106	.5557	.6894
23	.3605				
24	.3737	16	.6522	.6005	.7420

TABLE VIII

MORSE TWIST-DRILL AND STEEL-WIRE GAUGE

Gauge No.	Diameter Inch	Gauge No.	Diameter Inch	Gauge No.	Diameter Inch
1	.2280	28	.1405	55	.0520
2	.2210	29	.1360	56	.0465
3	.2130	30	.1285	57	.0430
4	.2090	31	.1200	58	.0420
5	.2055	32	.1160	59	.0410
6	.2040	33	.1130	60	.0400
7	.2010	34	.1110	61	.0390
8	.1990	35	.1100	62	.0380
9	.1960	36	.1065	63	.0370
10	.1935	37	.1040	64	.0360
11	.1910	38	.1015	65	.0350
12	.1890	39	.0995	66	.0330
13	.1850	40	.0980	67	.0320
14	.1820	41	.0960	68	.0310
15	.1800	42	.0935	69	.02925
16	.1770	43	.0890	70	.0280
17	.1730	44	.0860	71	.0260
18	.1695	45	.0820	72	.0250
19	.1660	46	.0810	73	.0240
20	.1610	47	.0785	74	.0225
21	.1590	48	.0760	75	.0210
22	.1570	49	.0730	76	.0200
23	.1540	50	.0700	77	.0180
24	.1520	51	.0670	78	.0160
25	.1495	52	.0635	79	.0145
26	.1470	53	.0595	80	.0135
27	.1440	54	.0550		

TABLE IX

DECIMAL EQUIVALENTS OF PARTS OF 1 INCH

Fraction	Decimal	Fraction	Decimal	Fraction	Decimal	Fraction	Decimal
	.0078		.2578		.5078		.7578
$\frac{1}{64}$	.0156	$\frac{17}{64}$	.2656	$\frac{33}{64}$	.5156	$\frac{49}{64}$	.7656
	.0235		.2735		.5235		.7735
$\frac{1}{32}$	.0313	$\frac{9}{32}$	.2813	$\frac{17}{32}$	.5313	$\frac{25}{32}$	.7813
	.0391		.2891		.5391		.7891
$\frac{3}{64}$	.0469	$\frac{19}{64}$	.2969	$\frac{35}{64}$	.5469	$\frac{51}{64}$	.7969
	.0547		.3047		.5547		.8047
$\frac{1}{16}$	.0625	$\frac{5}{16}$	.3125	$\frac{9}{16}$	.5625	$\frac{13}{16}$	.8125
	.0703		.3203		.5703		.8203
$\frac{5}{64}$	.0781	$\frac{21}{64}$	.3281	$\frac{37}{64}$	.5781	$\frac{53}{64}$	.8281
	.0860		.3360		.5860		.8360
$\frac{3}{32}$	.0938	$\frac{11}{32}$	.3438	$\frac{19}{32}$	.5938	$\frac{27}{32}$	.8438
	.1016		.3516		.6016		.8516
$\frac{7}{64}$	.1094	$\frac{23}{64}$	.3594	$\frac{39}{64}$	.6094	$\frac{55}{64}$	.8594
	.1172		.3672		.6172		.8672
$\frac{1}{8}$	.1250	$\frac{3}{8}$	.3750	$\frac{5}{8}$	.6250	$\frac{7}{8}$	.8750
	1328		.3828		.6328		.8828
$\frac{9}{64}$	.1406	$\frac{25}{64}$	.3906	$\frac{41}{64}$	.6406	$\frac{57}{64}$	.8906
	.1485		.3985		.6485		.8985
$\frac{5}{32}$	.1563	$\frac{13}{32}$	.4063	$\frac{21}{32}$	.6563	$\frac{29}{32}$	.9063
	.1641		.4141		.6641		.9141
$\frac{11}{64}$	.1719	$\frac{27}{64}$	.4219	$\frac{43}{64}$	.6719	$\frac{59}{64}$	.9219
	.1797		.4297		.6797		.9297
$\frac{3}{16}$	.1875	$\frac{7}{16}$	.4375	$\frac{11}{16}$	.6875	$\frac{15}{16}$	.9375
	.1953		.4453		.6953		.9453
$\frac{13}{64}$	.2031	$\frac{29}{64}$	.4531	$\frac{45}{64}$	.7031	$\frac{61}{64}$	.9531
	.2110		.4610		.7110		.9610
$\frac{7}{32}$	.2188	$\frac{15}{32}$	.4688	$\frac{23}{32}$	.7188	$\frac{31}{32}$	.9688
	.2266		.4766		.7266		.9766
$\frac{15}{64}$	.2344	$\frac{31}{64}$	.4844	$\frac{47}{64}$	.7344	$\frac{63}{64}$	.9844
	.2422		.4922		.7422		.9922
$\frac{1}{4}$	.2500	$\frac{1}{2}$	.5000	$\frac{3}{4}$	.7500	1	1.0000
	.2578		.5078		.7578		1.0078

TABLE X

DIMENSIONS OF PLATE WASHERS

Diameter of Bolt d Inch	Diameter of Washer D_2 Inches	Diameter of Hole Inches	Thickness of Washer t		Diameter of Bolt d Inches	Diameter of Washer D_2 Inches	Diameter of Hole Inches	Thickness of Washer t	
			Birmingham Wire Gauge No.	Approximate Value Inch				Birmingham Wire Gauge No.	Approximate Value Inch
$\frac{3}{16}$	$\frac{9}{16}$	$\frac{1}{4}$	18	$\frac{3}{64}$	$1\frac{1}{8}$	$2\frac{3}{4}$	$1\frac{1}{4}$	9	$\frac{5}{32}$
$\frac{1}{4}$	$\frac{3}{4}$	$\frac{5}{16}$	16	$\frac{1}{16}$	$1\frac{1}{4}$	3	$1\frac{3}{8}$	8	$\frac{11}{64}$
$\frac{5}{16}$	$\frac{7}{8}$	$\frac{3}{8}$	16	$\frac{1}{16}$	$1\frac{3}{8}$	$3\frac{1}{4}$	$1\frac{1}{2}$	8	$\frac{11}{64}$
$\frac{3}{8}$	1	$\frac{7}{16}$	14	$\frac{5}{64}$	$1\frac{1}{2}$	$3\frac{1}{2}$	$1\frac{5}{8}$	8	$\frac{11}{64}$
$\frac{7}{16}$	$1\frac{1}{4}$	$\frac{1}{2}$	14	$\frac{5}{64}$	$1\frac{5}{8}$	$3\frac{3}{4}$	$1\frac{3}{4}$	8	$\frac{11}{64}$
$\frac{1}{2}$	$1\frac{3}{8}$	$\frac{9}{16}$	12	$\frac{7}{64}$	$1\frac{3}{4}$	4	$1\frac{7}{8}$	8	$\frac{11}{64}$
$\frac{9}{16}$	$1\frac{1}{2}$	$\frac{5}{8}$	12	$\frac{7}{64}$	$1\frac{7}{8}$	$4\frac{1}{4}$	2	8	$\frac{11}{64}$
$\frac{5}{8}$	$1\frac{3}{4}$	$\frac{11}{16}$	10	$\frac{9}{64}$	2	$4\frac{1}{2}$	$2\frac{1}{8}$	8	$\frac{11}{64}$
$\frac{3}{4}$	2	$\frac{13}{16}$	10	$\frac{9}{64}$	$2\frac{1}{4}$	$4\frac{3}{4}$	$2\frac{3}{8}$	6	$\frac{13}{64}$
$\frac{7}{8}$	$2\frac{1}{4}$	$\frac{15}{16}$	9	$\frac{5}{32}$	$2\frac{1}{2}$	5	$2\frac{5}{8}$	5	$\frac{7}{64}$
1	$2\frac{1}{2}$	$1\frac{1}{16}$	9	$\frac{5}{32}$					

MACHINE DESIGN

(PART 2)

KEYS AND COTTERS

KEYS

1. **Keys** are iron or steel wedges used to secure wheels, cranks, and pulleys to shafts. It is the duty of the key to prevent the relative rotation of the pieces connected; if, for example, the pieces in question are a pulley and shaft, the function of the key is to prevent the pulley from turning on the shaft. Generally, the key will also prevent a wheel or pulley from moving lengthwise along the shaft.

FORMS OF KEYS

2. Keys may be divided into three classes: *sunk*, *concave*, and *flat*.

Sunk keys, by their resistance to shear, prevent motion between the shaft and the attached part. To this class

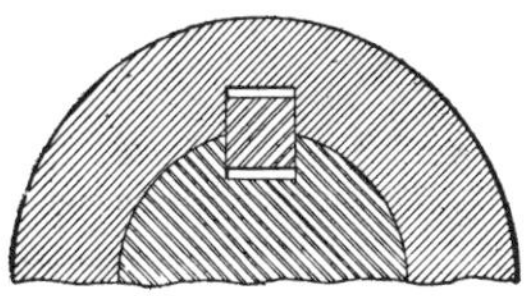

FIG. 1

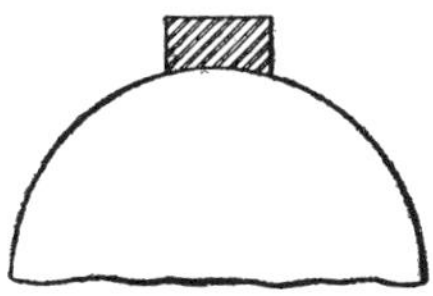

FIG. 2

belong keys that are fitted in slots cut both in the shaft and in the hub of the attached member. Fig. 1 shows a shaft with a sunk key.

Concave, or **saddle, keys** depend on the friction between them and the shaft to prevent relative rotation. In this case, a slot is provided only in the attached part, while the key is hollowed out to fit the shaft. Such keys are suitable for light work only. A concave key is shown in Fig. 2.

Flat keys use a combination of the means just mentioned to prevent the attached part from changing position. Also, in this key, only one slot is used, but instead of making the key to fit the shaft, a flat surface is cut on the latter to receive the key. This key is consequently more effective than the concave key. Fig. 3 shows a flat key, also called *key on the flat.*

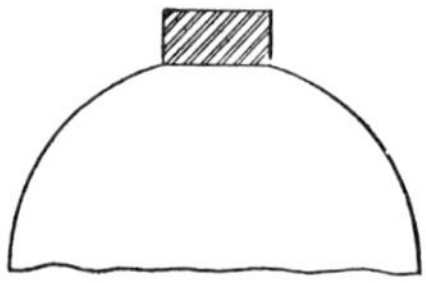

FIG. 3

The slot that is cut lengthwise in the shaft, or in the hub of the pulley, gear, etc., is called a **keyway,** or **key seat.**

3. There are two forms of sunk keys in use, each of which requires a different method for fitting the key in the key seat.

1. The **rectangular key,** shown in Fig. 4, is driven in light and usually fits at the top and bottom as well as at the sides. In order to insure a good fit and to make the driving in and the removal of the key less difficult, it is slightly tapered; that is, the top and bottom planes of the key are not parallel. As a key of this shape is liable to spring the parts out of true, it is not suitable for conditions where exact fitting is required; in general, it is used for fastening gear-wheels, pulleys, etc. to shafts.

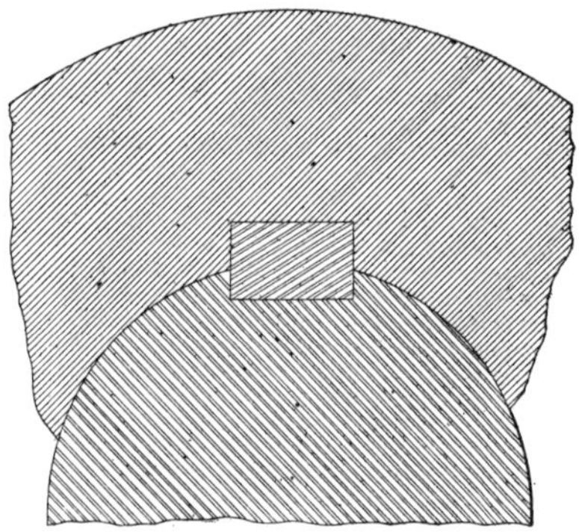

FIG. 4

NOTE.—To prevent any misunderstanding, the word *taper*, when used in this subject, will mean the gradual diminution in size of a slender object. Thus, should it be stated that a certain conical piece has a taper of 2 inches per foot, it would be meant that were the conical piece 1 foot long, the diameter at one end would be 2 inches larger than at the other.

The keys shown in Figs. 2 and 3 are also tapered in the manner just described. The taper varies from $\frac{3}{16}$ to $\frac{1}{8}$ or $\frac{3}{32}$ inch per foot, the smaller tapers being used in the most accurate work.

2. The **square key,** Fig. 1, fits tightly at the sides, but not at the top and bottom. The parts will therefore not be strained, which is an important consideration when accurate fitting is required, as in the case of machine tools. As a square key has no taper, it does not always prevent axial movement of the attached piece unless the latter is forced on the shaft. Otherwise, additional means, such as setscrews, end collars, etc., must be used for holding the piece in position.

4. In cases where a piece has to be adjusted on the shaft in various positions, a key seat of sufficient length is milled out in the shaft and a square key inserted. By the use of setscrews, which force the key toward the shaft, the piece may be held in any required position. It will thus be possible to move the piece without removing the key. The square key requires no taper.

5. Large wheels or pulleys may be fastened to the shaft by using two, three, or four keys. Fig. 5 shows a method

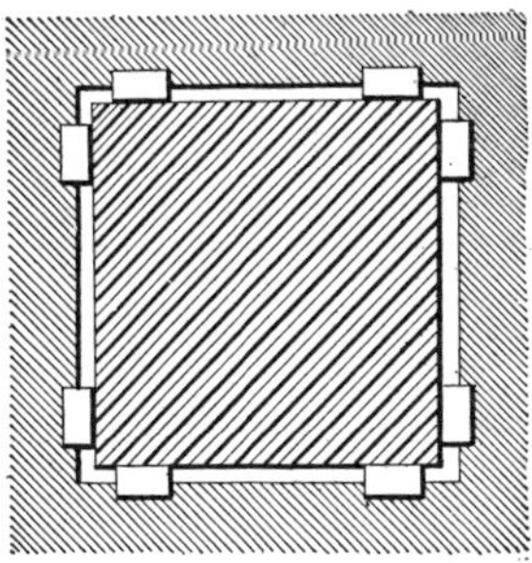

Fig. 5

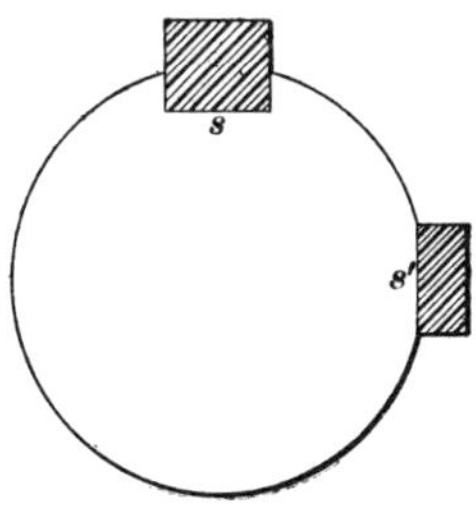

Fig. 6

of keying a piece to a square shaft. When several keys are thus used the wheel or pulley may be centered on the shaft by means of the keys.

If a pulley is accidentally bored a little too large for the shaft, it may be prevented from rocking by using both a sunk key and a flat key, as shown in Fig. 6. The flat key *s'* is placed at a distance around the shaft of about 90° from the sunk key *s*, and the pulley is thus supported at three points.

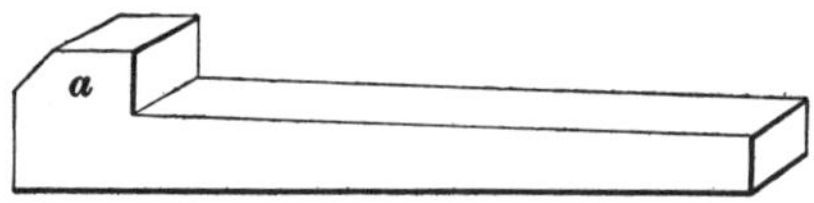

FIG. 7

When a key cannot be conveniently driven out from the small end, it should be made with a gib head ***a***, as shown in Fig. 7. The head forms a shoulder to drive against.

6. Sliding, or feather, keys, or splines, are used to prevent a piece from rotating on a shaft, and at the same time allow it to slide lengthwise. The key, or spline, is usually fastened to the piece, and is free to slide in the keyway of the shaft, though the operation is sometimes reversed and the key is fastened to the shaft. This form of key is made without any taper. Various methods of fixing the key to the wheel or pulley are shown in Figs. 8, 9, and 10. In Fig. 8, the key is dovetailed and driven tightly into the hub *m*.

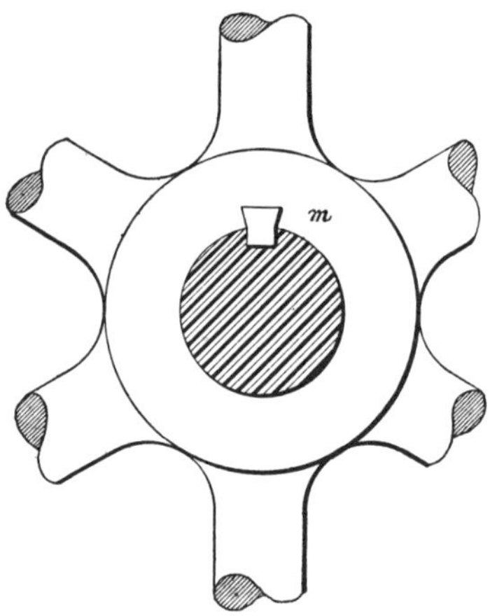

FIG. 8

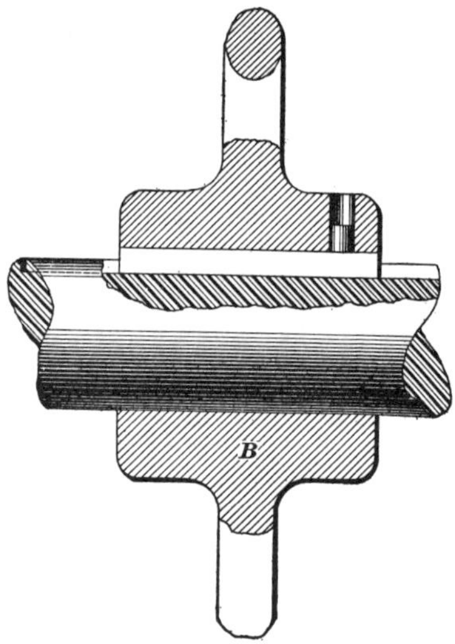

FIG. 9

When the hub of a wheel in which there is a feather abuts against a collar or a bearing, it is evident that the feather must not project. In such a case the feather key, Fig. 8,

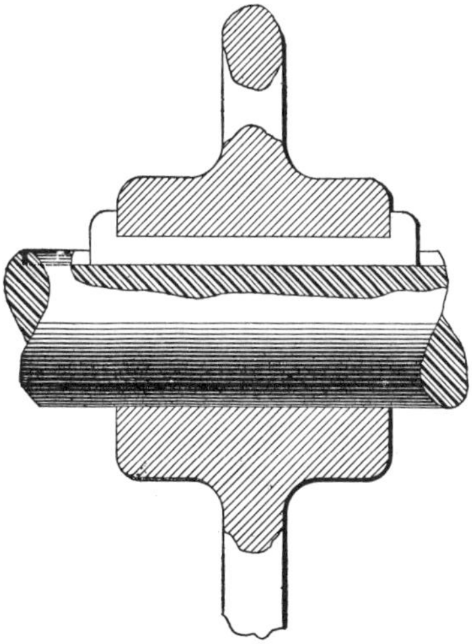

FIG. 10

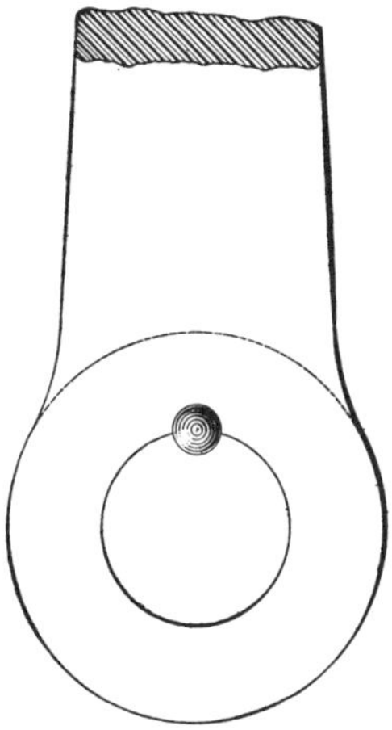

FIG. 11

or the flush feather key, Fig. 9, may be used; otherwise, the key may have gib heads, as shown in Fig. 10.

7. Round, or **pin, keys** may be used when the piece is shrunk on to the shaft, as for example, a small crank, as shown in Fig. 11. A hole is drilled partly in the shaft and partly in the crank, and a round pin is driven into the hole, as shown in the figure.

8. Woodruff Key.—The **Woodruff key,** shown in Fig. 12, is used to some extent on machine tools. This key is

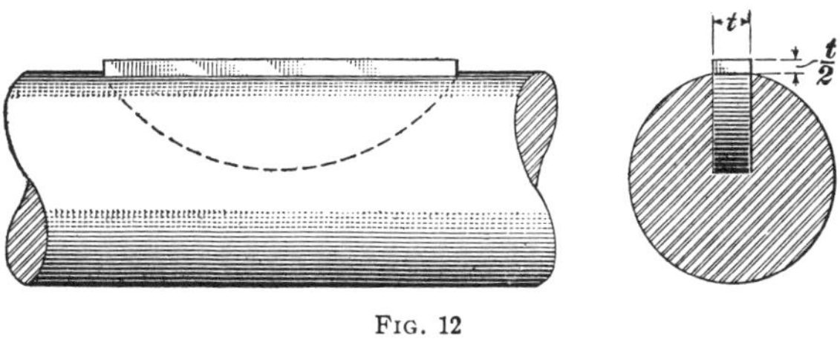

FIG. 12

nearly semicircular in outline and rests in a seat, the bottom of which is curved in conformity with the key. This arrangement will allow the key to adjust itself by a partial rotation until a perfect contact is obtained between its flat edge and the key seat in the piece to be fastened. By reason of its form, the lower key seat is readily made by a milling cutter having a diameter equal to that of the curved portion of the key, thus effecting a great saving of labor in making the seat and in fitting the key.

As shown in Fig. 12, the key is of great depth relative to its width, and is therefore amply insured against tipping. The amount that the key projects above the shaft is equal to $\frac{1}{2}t$, t being the thickness of the key. If the shearing resistance of one key is insufficient, more keys may be added by arranging them lengthwise.

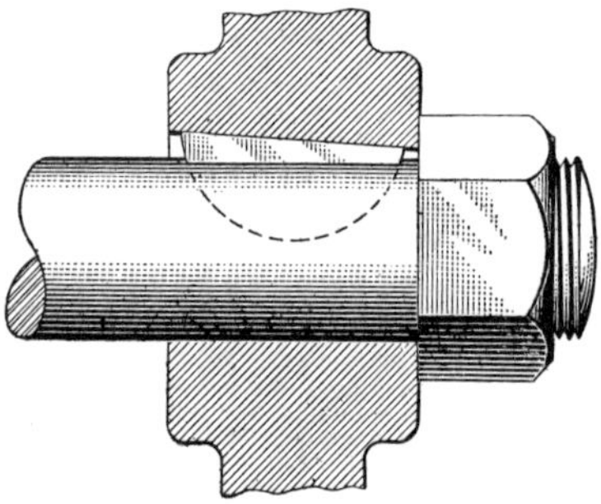

FIG. 13

Fig. 13 shows a hub in which the flat key seat inclines toward the axis of the hub. In this case, while the hub is being placed in position, the key will adjust itself by a partial rotation until it bears against the hub throughout the length of its flat edge.

STRENGTH AND PROPORTIONS OF KEYS

9. A sunk key is subjected to two kinds of stresses. The twisting of the piece on the shaft tends to shear the key, and also to crush it by compression.

Let b = width of key, in inches;
t = thickness of key, in inches;
l = length of key, in inches;
S_2 = safe shearing stress allowable, in pounds per square inch;
S_3 = safe crushing stress allowable, in pounds per square inch;
P = force, in pounds, acting at rim of wheel or pulley;

R = radius of wheel or pulley, in inches;
d = diameter of shaft, in inches.

The shearing area of the key is $b\,l$; hence, the safe resistance of the key to shearing is $b\,l\,S_2$. Taking moments about the center of the shaft,

$$b\,l\,S_2 \times \tfrac{1}{2} d = P \times R$$

or

$$b\,l = \frac{2\,P\,R}{S_2\,d} \qquad (1)$$

Supposing the key to be half bedded in the shaft, the crushing area is $\frac{1}{2}\,t\,l$, and the resistance to crushing is $\frac{1}{2}\,t\,l\,S_3$. If the key is designed to be equally strong against shearing and crushing, the shearing resistance must equal the crushing resistance, or

$$b\,l\,S_2 = \tfrac{1}{2}\,t\,l\,S_3$$

or

$$b = \tfrac{1}{2}\,t\frac{S_3}{S_2} \qquad (2)$$

If, now, the crushing strength of the material is assumed to be double the shearing strength, $\frac{S_3}{S_2} = 2$, and $b = t$. With square and feather keys, S_3 may be nearly equal to $2\,S_2$; but for flat, tapered keys the crushing action is smaller. The reason for this difference is that the key, in its tendency to tip, will act as a strut and thus exert a certain radial bursting pressure on the hub. This bursting pressure will to some extent reduce the side pressure on the key, and, as the danger of crushing is smaller, a smaller factor of safety may be used. In any case, b is rarely less than t, and is frequently made greater.

For shearing, a factor of safety of about 10 may be used, giving a safe shearing stress S_2 of 5,000 pounds per square inch for wrought iron, and 7,000 pounds for steel. Then, formula **1** becomes

$$\left.\begin{aligned} b\,l &= \frac{P\,R}{2{,}500\,d} \text{ for wrought iron} \\ b\,l &= \frac{P\,R}{3{,}500\,d} \text{ for steel} \end{aligned}\right\} \qquad (3)$$

Instead of the twisting moment $P\,R$ of formula **3,** it may be more convenient to use the horsepower transmitted by the shaft and also its number of revolutions.

Let N = number of revolutions per minute;
H = horsepower.

Then, a point on the circumference of the wheel or pulley moves $2\pi R N$ inches, or $\frac{2\pi R N}{12}$ feet, per minute.

Hence, if a force P constantly acts at the circumference of the wheel, the work done per minute is

$$\frac{2\pi R N}{12} \times P \text{ foot-pounds}$$

Therefore, $\frac{2\pi R N P}{12} \div 33{,}000 = H$, the horsepower.

$$\frac{12 \times 33{,}000 \times H}{2\pi N} = PR = 63{,}025 \frac{H}{N} \qquad \textbf{(4)}$$

Substituting the value of PR from formula **4**, in formula **3**,

$$\left.\begin{aligned} b\,l &= \frac{25\,H}{d\,N} \text{ for wrought iron} \\ b\,l &= \frac{18\,H}{d\,N} \text{ for steel} \end{aligned}\right\} \qquad \textbf{(5)}$$

Formulas **3** and **5** may be used in calculating the sizes of keys for large work. For small shafts, the sizes given by formulas **3** and **5** are much smaller than are used in actual practice.

10. Designers frequently adopt some standard ratio between the depth and the width of the key, the ratio varying from $\frac{1}{2}$ to $\frac{5}{6}$. In this Section, the ratio $\frac{2}{3}$ is adopted; that is. $t = \frac{2}{3} b$.

EXAMPLE.—The maximum pressure on an engine crankpin is 12,500 pounds, and the length of crank is 10 inches. Assume the diameter of the shaft to be 5 inches, and the length of the key the same. What should be the dimensions of a wrought-iron key to hold the crank to the shaft? The crank is not to be shrunk on the shaft.

SOLUTION.—Transposing formula **3**,

$$b = \frac{PR}{2{,}500\,d\,l} = \frac{12{,}500 \times 10}{2{,}500 \times 5 \times 5} = 2 \text{ in. Ans.}$$

$$t = \tfrac{2}{3} b = \tfrac{2}{3} \times 2 = 1\tfrac{1}{3}, \text{ say } 1\tfrac{11}{32}, \text{ in. Ans.}$$

11. In ordinary designing, the sizes of keys are determined by empirical formulas that give an excess of strength. For ordinary keys in cranks, driving pulleys, etc. the following proportion may be adopted:

$$\left.\begin{aligned} b &= .21\,d + .04 \text{ inch} \\ t &= .12\,d + .04 \text{ inch} \end{aligned}\right\} \qquad (1)$$

When d is less than $1\frac{5}{16}$ inches, then

$$\left.\begin{aligned} b &= \frac{d}{3} \\ t &= \frac{d}{5} \end{aligned}\right\} \qquad (2)$$

Using formula **1** in the example of the crank-shaft, Art **10**,

$$b = .21\,d + .04 = .21 \times 5 + .04 = 1\tfrac{1}{8} \text{ inches, nearly}$$
$$t = .12\,d + .04 = .12 \times 5 + .04 = \tfrac{5}{8} \text{ inch, nearly}$$

Unless otherwise stated, formula **1** may be used in solving problems.

As an aid in ordinary designing, Table I, at the end of this Section, may be used. This table gives dimensions closely approximating those formed by formulas **1** and **2.**

Table II, also at the end of this Section, gives the dimensions of square keys, as used by William Sellers & Company. In this case, the keys are supposed to fit tightly sidewise, but not necessarily so on top and bottom. On the other hand, the hub is supposed to fit very tightly on the shaft.

For sliding feather keys, the following formulas give the proportions used in ordinary practice:

$$\left.\begin{aligned} b &= \frac{d}{8} + \tfrac{1}{16} \text{ inch} \\ t &= \frac{3\,d}{16} + \tfrac{1}{16} \text{ inch} \end{aligned}\right\} \qquad (3)$$

12. In some instances, pulleys may be keyed to a large shaft and yet transmit a small amount of power. In such cases the dimensions of a key, if based on the actual diameter of the shaft, will be much too large. The proportions may therefore be based on the diameter of a shaft that would be necessary to transmit the power of the pulley in question, and no more. Letting H represent the horsepower transmitted by the pulley, and N the revolutions per minute of the shaft,

$$d = 5\sqrt[3]{\frac{H}{N}}$$

and this value of d may be used in formula **2**, Art. **11.**

EXAMPLE.—A pulley transmitting 2 horsepower is keyed to a shaft 4 inches in diameter and making 120 revolutions per minute. Determine the dimensions of the key.

SOLUTION.—The diameter of a shaft to transmit 2 H. P. is

$$d = 5\sqrt[3]{\frac{H}{N}} = 5\sqrt[3]{\frac{2}{120}} = 1.28 \text{ in. Ans.}$$

From formula **2**, Art. **11**,

$$b = \frac{1.28}{3} = \tfrac{7}{16} \text{ in. Ans.}$$

$$t = \frac{1.28}{5} = \tfrac{1}{4} \text{ in. Ans.}$$

EXAMPLES FOR PRACTICE

1. Calculate the dimensions of an ordinary sunk key for a shaft $3\frac{1}{2}$ inches in diameter. Ans. $\frac{1}{2}$ in. $\times$ $\frac{3}{4}$ in.

2. Calculate the dimensions of a feather key for a shaft $2\frac{3}{4}$ inches in diameter. Ans. $\frac{13}{32}$ in. $\times$ $\frac{9}{16}$ in.

3. A wrought-iron key is used to fasten a flywheel on a 6-inch shaft. If the maximum pressure on the crankpin is 15,000 pounds, and the

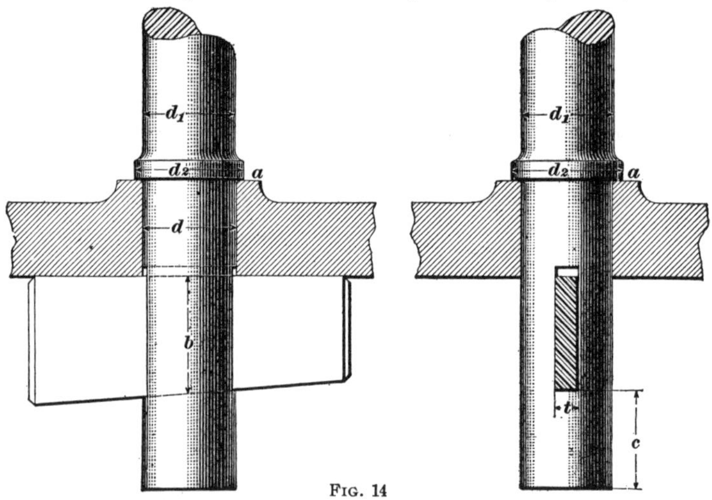

FIG. 14

crank radius is 16 inches, what should be the dimensions of the key, its length being 8 inches? Ans. 2 in. $\times$ $1\frac{11}{32}$ in.

4. A pulley transmits $3\frac{1}{2}$ horsepower, and is keyed to a shaft 5 inches in diameter and making 150 revolutions per minute. Calculate the dimensions of the key. Ans. $\frac{1}{2}$ in. $\times$ $\frac{5}{16}$ in.

COTTERS

13. A **cotter** is an iron or steel bar that is driven through one or both of two pieces to be connected, and holds them together by its resistance to shearing at two *transverse* cross-sections. A simple form of a cotter is shown in Fig. 14. The cotter passes through the rod only, and acts when the rod is in tension. The enlargement, or collar, *a* on the rod prevents any downward movement, and therefore resists thrust.

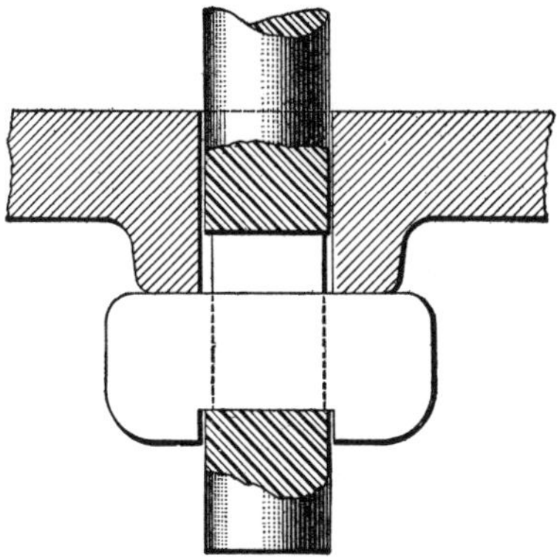

FIG. 15

Fig. 15 shows a cotter with gib ends. Since in this case the rod is not provided with a collar, this arrangement will resist tension only.

In the arrangement shown in Fig. 16, the cotter is divided into two parts, the one with hooked ends being called the

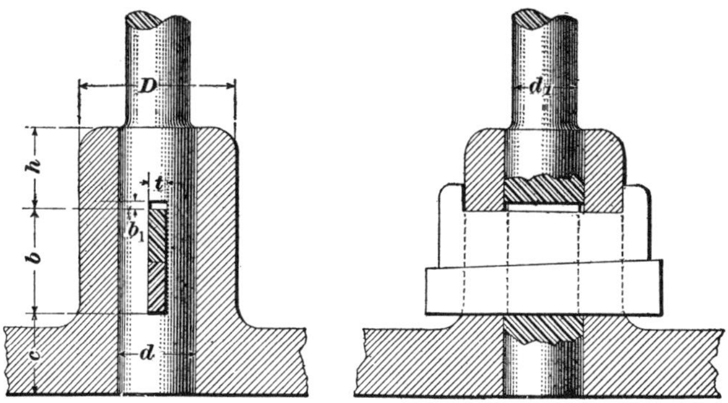

FIG. 16

gib and the other the cotter. In this construction, the rod should be placed in tension only. If it is necessary to provide for thrust, the end of the rod should be tapered as shown in Fig. 17.

Fig. 18 shows an arrangement in which a rod is cottered into a socket. As shown in Figs. 16 and 18, the cotter is long and tapered; it serves, therefore, as a means of adjusting the length of the connected pieces. By driving the cotter farther in, the total length of the two pieces is lessened, and vice versa.

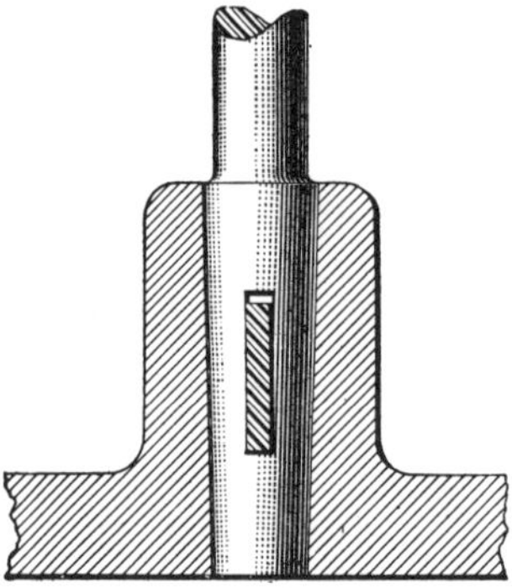

Fig. 17

NOTE.—The dimensions b and c, Fig. 18, run to the intersection of the slant side of the cotter and the center line of the rod.

A cotter may be used to connect two straps m and n to a rod l, as shown in Fig. 19. When driven down, the friction between the cotter and the lower strap may cause the latter to open out, as shown by the dotted lines; hence, it is desirable in

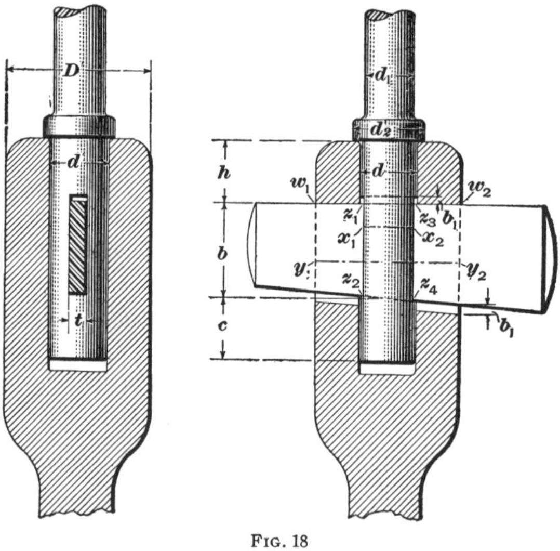

Fig. 18

such a case to use a cotter combined with a gib, as shown in Fig. 20, or with two gibs, as shown in Fig. 21.

The gibs serve to keep the straps from spreading. In Fig. 22, the side ab of the gib and cd of the cotter are parallel to each other and perpendicular to the axis of the rod, the taper coming between the gib and the cotter.

14. Strength and Proportions of Cotters.—In designing a cotter connection, the following points should be taken into account. Referring to the illustration of the cotter, Fig. 14, (1) the cross-section $b\,t$ of the cotter must be sufficient to withstand the shearing force; (2) the thickness t must be great enough to provide against failure by crushing;

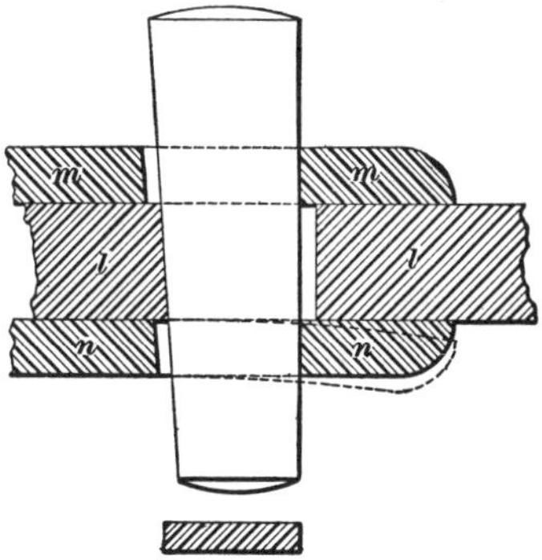

FIG. 19

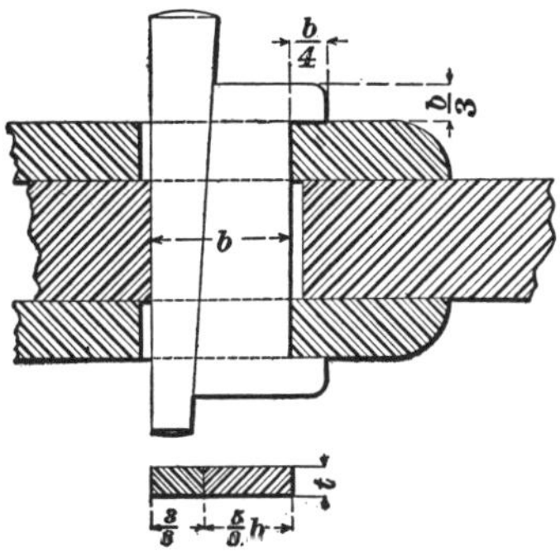

FIG. 20

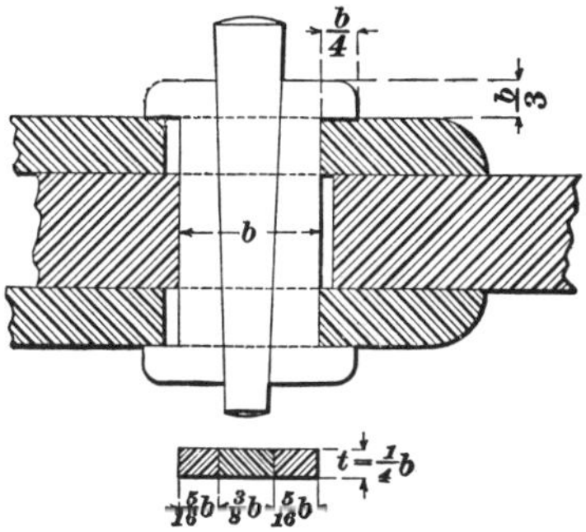

FIG. 21

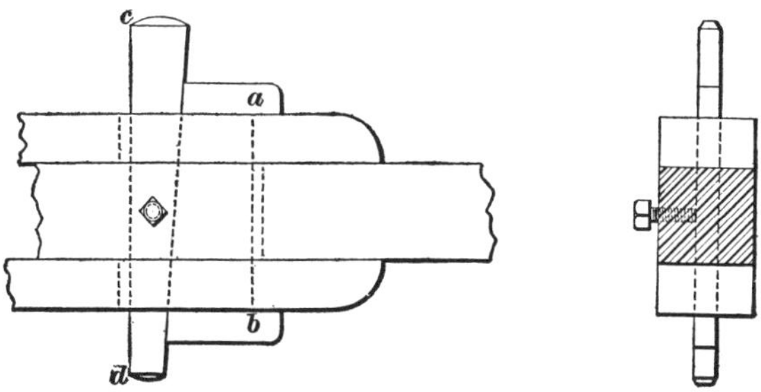

FIG. 22

and (3) the two diameters d_1 and d should be so designed that the rod is of uniform strength throughout.

Let P = force, in pounds, exerted on the rod;
S_1 = safe tensile strength of rod, in pounds per square inch;
S_2 = safe shearing strength of cotter, in pounds per square inch;
S_3 = safe compressive strength of cotter or rod, in pounds per square inch.

The various diameters and other dimensions are indicated in the figures.

Consider the arrangement shown in Fig. 18, and conceive a transverse section taken through the cotter hole. The net area of the rod is $.7854\,d^2 - d\,t$, very nearly; the shearing area of the cotter is $2\,b\,t$; the area of the cotter subject to crushing against the rod is $d\,t$, and against the socket $(D - d)\,t$, very nearly. The area of the socket subject to tension is $.7854\,(D^2 - d^2) - (D - d)\,t$, and the area of the smaller part of the rod is $.7854\,d_1^2$. Finally, the cotter may be considered as a beam uniformly loaded, in which case the bending moment will be $\frac{P\,d}{8} = \frac{I}{c}S_1$. From the table in *Strength of Materials*, Part 2, $I = \frac{1}{12}\,b\,d^3$ and $c = \frac{1}{2}\,d$. In the moment-of-inertia formula, b and d represent the dimensions t and b, respectively, of the key. Hence, $\frac{P\,d}{8} = \frac{t\,b^2}{6}S_1$, and solving for P, then

$$P = \tfrac{4}{3}\frac{t\,b^2}{d}S_1 \qquad (a)$$

The other formulas representing the tensile, shearing, and crushing stresses are as follows:

At the section $x_1\,x_2$, Fig. 18, for tension in the rod,

$$P = (.7854\,d^2 - d\,t)\,S_1 \qquad (b)$$

At the section $y_1\,y_2$, for tension in the socket,

$$P = [.7854\,(D^2 - d^2) - (D - d)\,t]\,S_1 \qquad (c)$$

Considering the tension in the rod of diameter d_1,

$$P = .7854\,d_1^2\,S_1 \qquad (d)$$

At the sections $z_1\,z_2$ and $z_3\,z_4$, for shearing stresses in the cotter,

$$P = 2\,b\,t\,S_2 \qquad (e)$$

At the surface $z_2 z_4$, for crushing stresses between the rod and the cotter,

$$P = d\,t\,S_3 \qquad (f)$$

At the surfaces $w_1 z_1$ and $z_3 w_2$, for crushing stresses between the socket and the cotter,

$$P = (D - d)\,t\,S_3 \qquad (g)$$

It is supposed that the rod and the cotter are made of the same material, either wrought iron or steel. Taking $\frac{S_2}{S_1} = \frac{4}{5}$, as in *Machine Design*, Part 1,

$$S_2 = \tfrac{4}{5} S_1$$

Substituting this value of S_2 in formula (e),

$$P = 2\,b\,t\,\tfrac{4}{5} S_1 = 1.6\,b\,t\,S_1$$

Placing this value of P equal to that found by means of formula (a),

$$1.6\,b\,t\,S_1 = \tfrac{4}{3}\frac{t\,b^2}{d} S_1$$

whence $$b = 1.2\,d$$

If this value of b is inserted in formula (e) and the values of P in formulas (e) and (b) are then equated,

$$2 \times 1.2\,d\,t\,S_2 = (.7854\,d^2 - d\,t)\,S_1$$

Inserting for S_2 the value $\frac{4}{5} S_1$,

$$2.4\,d\,t\,\tfrac{4}{5} S_1 = (.7854\,d^2 - d\,t)\,S_1$$

$$1.92\,d\,t = .7854\,d^2 - d\,t$$

$$t = \frac{.7854\,d}{2.92} = .27\,d, \text{ or } \tfrac{1}{4}\,d, \text{ nearly}$$

To prevent failure from crushing by the cotter, the bearing surface of the socket should equal that of the rod. Equating the values of P in formulas (g) and (f),

$$(D - d)\,t\,S_3 = d\,t\,S_3$$

Then, $$D = 2\,d$$

Equating the values of P in formulas (b) and (f),

$$(.7854\,d^2 - d\,t)\,S_1 = d\,t\,S_3,$$

and making $t = \frac{1}{4}\,d$,

$$\frac{S_3}{S_1} = 2.14, \text{ or } 2, \text{ approximately}$$

Experience has shown that this ratio may be found where materials and workmanship are good.

Equating the values of P in formulas (b) and (d), and taking $t = \frac{1}{4}d$,

$$.7854\,d^2 - d\,t = .7854\,d_1^2$$

or

$$.7854\,d^2 - \frac{d^2}{4} = .7854\,d_1^2$$

and

$$d_1 = .83\,d$$

The diameter d_2 of the collar, Fig. 18, should be such that its bearing surface is at least equal to that of the cotter in the rod. Hence,

$$.7854\,(d_2^2 - d^2) = d\,t = \frac{d^2}{4}$$

and

$$.7854\,d_2^2 = 1.0354\,d^2$$

$$d_2 = d\sqrt{\frac{1.0354}{.7854}} = 1.15\,d$$

A clearance b_1 is shown in Figs. 16 and 18 between the cotter and the socket, and the cotter and the rod. The amount of this clearance depends on the taper of the cotter and on the distance through which the latter may be moved. In general, this clearance is not less than $\frac{1}{8}$ inch.

Collecting the foregoing results,

$b = 1.2\,d$	$d_2 = 1.15\,d$
$t = \frac{1}{4}\,d$	$h = c = \frac{3}{4}\,d$ to $1\frac{1}{4}\,d$
$D = 2\,d$	$b_1 = \frac{1}{8}$ inch, or more
$d_1 = .83\,d$	

15. If a steel cotter is used in a wrought-iron rod, b may be made equal to d, the other dimensions remaining the same as just given.

For a cotter of the form shown in Fig. 20 or 21, it is good practice to make $t = \frac{1}{4}\,b$, and $b\,t = \frac{5}{4} \times$ sectional area of strap. The width b is the same whether a single cotter, a gib and cotter, or two gibs and cotter are used. The other proportions are shown in the figures.

EXAMPLE.—Suppose that in Fig. 22 the strap is $\frac{7}{8}$ in. $\times\ 3\frac{1}{2}$ in.; find the thickness of the cotter, and the combined width of gibs and cotter.

SOLUTION.— $b\,t = \frac{5}{4} \times \frac{7}{8} \times 3\frac{1}{2} = \frac{245}{64}$. Making $t = \frac{1}{4}\,b$, as just stated, $b\,t = \frac{1}{4}\,b^2 = \frac{245}{64}$; whence

$$b = \sqrt{\frac{245}{16}} = 3.91 = 3\tfrac{15}{16} \text{ in. Ans.}$$

$$t = \tfrac{1}{4}\,b = 1 \text{ in., nearly. Ans.}$$

The width of cotter is, then, $\frac{3}{8}b$, or $1\frac{1}{2}$ inches, and the width of each gib is $\frac{5}{16}b$, or $1\frac{7}{32}$ inches, according to the dimensions given in Fig. 21.

16. Taper of Cotters.—The greatest allowable taper that a cotter may have without danger of slacking back is about $1\frac{3}{4}$ inches per foot. Usually, when the cotter is not secured, the taper is from $\frac{1}{4}$ to $\frac{1}{2}$ inch per foot. If fastened by a setscrew or bolt and nut, as in Figs. 22 and 23, the cotter may have a taper of $1\frac{1}{2}$ to 2 inches per foot. The taper per foot of a cotter is found by means of proportion from its increase in width. For instance, if a cotter is 2 inches wide at one end, $2\frac{3}{8}$ inches wide at the other, and 9 inches long, the increase in width is $2\frac{3}{8} - 2 = \frac{3}{8}$ inch. Then, if x is the taper per foot, the proportion $\frac{3}{8} : x = 9 : 12$, will give the value of x. Thus,

$$x = \frac{3}{8} \times \frac{12}{9} = \frac{1}{2} \text{ inch per foot}$$

17. Locking Arrangements.—In Figs. 22 and 23 are shown **locking arrangements for cotters.** In Fig. 22, the cotter is held by a setscrew, the point of which fits into a groove cut into the cotter. The diameter of the setscrew may be $\frac{1}{3}b + \frac{1}{8}$ inch. In the arrangement shown in Fig. 23, the end of the cotter is a screw, and the cotter is secured by a nut on an extra seat. This method is used where the cotter has an excessive taper.

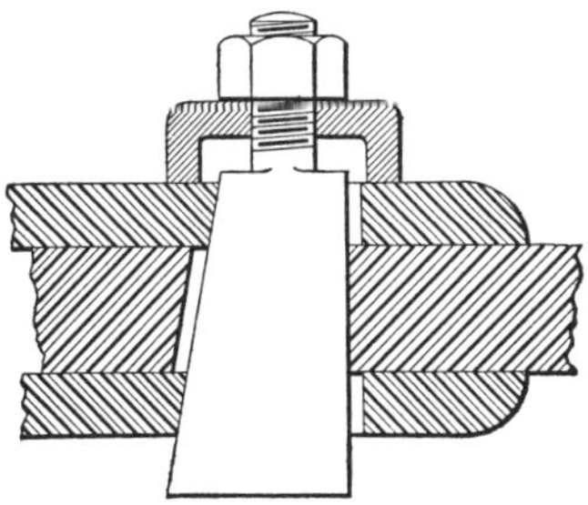

Fig. 23

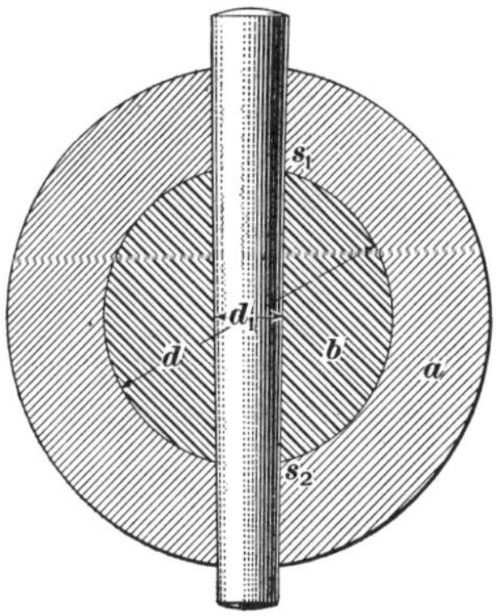

Fig. 24

18. Taper Pins.—A **taper pin** may be classed either as a key or as a cotter, but as its action is more like that of a cotter it will be described at this time. As shown in Fig. 24, the taper pin passes through both the hub *a* and the shaft *b* and is given a certain amount of taper to insure a good fit and to facilitate its removal. This pin is used mostly for light work, as it cannot be made large enough in diameter to resist a relatively large amount of shear without weakening the cross-sectional area of the shaft. In many cases a taper pin is inserted more for the purpose of preventing motion of the hub along the shaft than for the purpose of driving.

Let P = force, in pounds, acting at rim of wheel or pulley;
R = radius of wheel or pulley, in inches;
d = diameter of shaft, in inches;
d_1 = average diameter of pin, in inches;
S_2 = safe shearing stress.

As the shear will take place at the points s_1 and s_2, Fig. 24, there will be two shearing areas. The total shearing area of the pin, then, is $2 \times .7854\, d_1^2$, and the safe resistance to shearing is $2 \times .7854\, d_1^2 S_2$. Taking moments about the center of the shaft,

$$2 \times .7854\, d_1^2 S_2 \frac{d}{2} = PR$$

and

$$d_1 = 1.13\sqrt{\frac{PR}{d S_2}}$$

The pins are tapered according to certain adopted standards corresponding to the Morse standard reamers. Table III, at the end of this Section, gives the standard diameters of the pins, the length varying from $\frac{3}{4}$ inch to 6 inches. The taper is $\frac{1}{4}$ inch per foot.

EXAMPLE.—The hub of a lever 10 inches long is to be fastened by means of a taper pin to a shaft 2 inches in diameter. If the end of the lever is to support a weight of 180 pounds and the pin is made of steel having a safe fiber stress S_2 of 7,000 pounds, what will be the diameter of the pin?

SOLUTION.—From the formula of this article, $d_1 = 1.13\sqrt{\frac{180 \times 10}{2 \times 7{,}000}}$ = .405 in., corresponding to a No. 7 taper pin. Ans.

EXAMPLES FOR PRACTICE

1. Find the dimensions of a rod and socket of the form shown in Fig. 18, assuming that S_1 is equal to 6,000 pounds. The load or pull on the rod is 4,600 pounds.

$$\text{Ans.}\begin{cases} d = 1\tfrac{3}{16} \text{ in.}; & b = 1\tfrac{1}{2} \text{ in.};\\ d_1 = 1 \text{ in.}; & t = \tfrac{19}{64} \text{ in.};\\ d_2 = 1\tfrac{3}{8} \text{ in.}; & D = 2\tfrac{3}{8} \text{ in.}\\ h = 1 \text{ in.}; & \end{cases}$$

2. A cotter of the form shown in Fig. 15 resists a pull of 3,200 pounds. Find the necessary breadth and thickness, assuming that S_2 is 4,000 pounds and that the thickness is one-fourth the breadth.

$$\text{Ans.}\begin{cases} b = 1.27 = 1\tfrac{1}{4} \text{ in.}\\ t = \tfrac{5}{16} \text{ in.} \end{cases}$$

3. A cotter and two gibs connect two straps to a rod, as shown in Fig. 21. Supposing the pull on the rod to be 9,000 pounds and taking $S_2 = 5{,}400$ for steel, find the dimensions of cotter and gibs.

$$\text{Ans.}\begin{cases} t = \tfrac{15}{32} \text{ in.}\\ \text{Width of cotter} = \tfrac{11}{16} \text{ in.}\\ \text{Width of gibs} = \tfrac{9}{16} \text{ in.} \end{cases}$$

4. Calculate the dimensions of a steel cotter that fastens a wrought-iron rod $2\frac{3}{8}$ inches in diameter. Ans. $2\frac{3}{8}$ in. $\times \frac{19}{32}$ in.

5. A cotter is $1\frac{3}{4}$ inches wide in the middle and tapers on each side. If the cotter is 18 inches long, what is its width at each end? Assume that the taper is $\frac{1}{2}$ inch to the foot. Ans. $2\frac{1}{8}$ in. and $1\frac{3}{8}$ in.

RIVETED JOINTS

RIVETS

19. Distinctive Features.—Rivets differ from bolts in the following respects: (1) Instead of a detachable nut, the rivet has a supplementary head, which is formed from the body of the rivet after it has been inserted in its proper place. (2) The diameter of the rivet body is increased in size after the rivet is placed in position and while the new head is being formed.

As rivets are provided with two heads, they are used to unite permanently two or more parts, mostly steel or iron plates. As the safe union of the combined parts in this case does not depend on the strength of screw threads, as in the case of bolts and nuts, but on the cross-sectional area of the rivet body, some metals that would be less serviceable for bolts are available for rivets. Among these metals may be mentioned copper, brass, and aluminum, although steel and wrought iron are most commonly used.

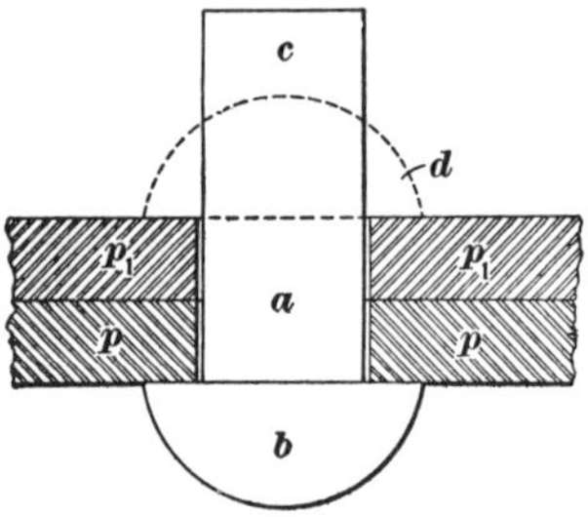

FIG. 25

20. Elements of a Rivet. A rivet in the form furnished by the manufacturer, that is, before being provided with a second head, is termed a **blank.** In Fig. 25 is shown a blank, in which a is the **shank,** b the **head,** and c the **point.** The head, which is made subsequently from the point and is shown by dotted lines at d, is termed the **point head,** or, simply, **point.** As will be noticed, the shank does not fit the hole tightly; this difference in diameter between the hole and the shank generally amounts to $\frac{1}{16}$ inch. The shank is slightly tapered toward the point. To form the point head, the part of the shank projecting beyond the plate p_1 must have sufficient material to form not only the point head d, but also to provide the additional material required for the shank itself, as the diameter of the latter increases to that of the hole. In general, the protruding part of the shank is of

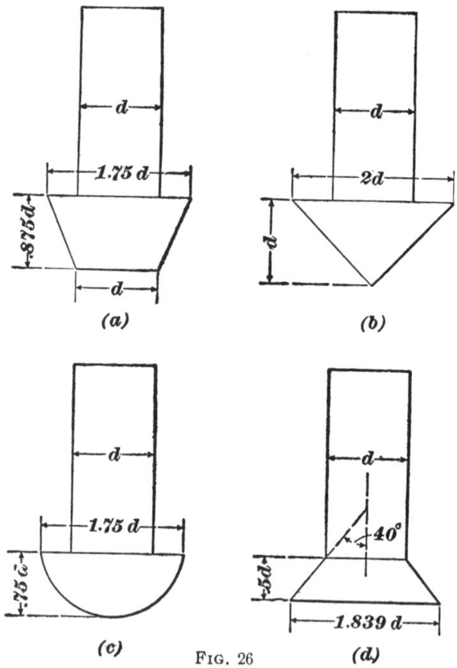

FIG. 26

a length amounting to from 1.3 to 1.7 times the diameter of shank.

21. Forms of Rivet Heads.—The forms of rivet heads in general use are shown in Fig. 26. The one shown at (*a*) is known as the *conical head;* at (*b*), the *steeple head;* at (*c*). the *button head;* and at (*d*), the *countersunk head.*

The proportions ordinarily given to these rivet heads are indicated in the various views in terms of the diameter *d*; but these dimensions are not always adopted. The countersunk head shows the most variation from the proportions given, both in regard to the height and to the diameter of head. Consequently, the angle between the side of the head and the axis of the rivet will also vary, and may be as small as 30° instead of 40°, as shown.

22. Rivet Holes.—The holes for rivets may be either drilled or punched, the former method being adopted where accuracy and strength are required. Accuracy cannot always be insured by punching, partly because the spacing of the holes may vary and also because the holes themselves are not truly cylindrical, but conical. The reason for this inaccuracy is that the diameter of the hole in the die is larger than that of the punch; hence, the hole in the plate has a larger diameter at the bottom than at the top. Punching is also believed to injure the plates, particularly if they are at all hard or brittle. Another objection to punching is that a ragged edge is produced, which has a tendency to decrease the shearing resistance of the rivet. Many makers, therefore, prefer to punch the hole smaller than its intended diameter and then ream it out. This operation cuts away the injured metal around the hole and makes it clean-cut and cylindrical. Annealing after punching will partly remove the injury caused by punching, but the better practice, where the greatest strength is required, is to punch the hole small and to ream to size.

23. Riveting.—The riveting, that is, the formation of the point head, may be done either by hand or by machine; in modern practice, the latter method is used to a greater extent. In either case the rivet is heated to a red heat and then inserted in the hole prepared for it. When the riveting is done by machine, the rivet is subjected to the pressure of two dies having the shape, or form, that the point head is to receive. As a result of the hammering by hand or of the pressure of the machine, the shank swells out and completely fills the hole. It is advantageous to have the shank hotter at the head end, thereby insuring that the swelling is complete before the point head is formed. When the rivet cools and contracts, the two heads are drawn firmly against the intervening plates.

In riveting, the rivet holes are often brought into line by means of a *drift pin*, which is a long, round taper pin that is small enough at one end to enter two rivet holes not well matched, and large enough at the other end to bring the holes into alinement. In boiler construction, the use of drift pins for enlarging the holes is prohibited by the best authorities.

24. Examples of Riveting.—In Figs. 27, 28, 29, 30, and 31 are shown some examples of the proportions of rivet shanks and the finished point heads. The dimensions of some of the heads differ from those given in Fig. 26, for the reason that there is no uniformly adopted standard, the rivet forms being left, to some extent, to the preference of the various manufacturers. The original form of the shank is shown by dotted lines and its dimensions are also indicated.

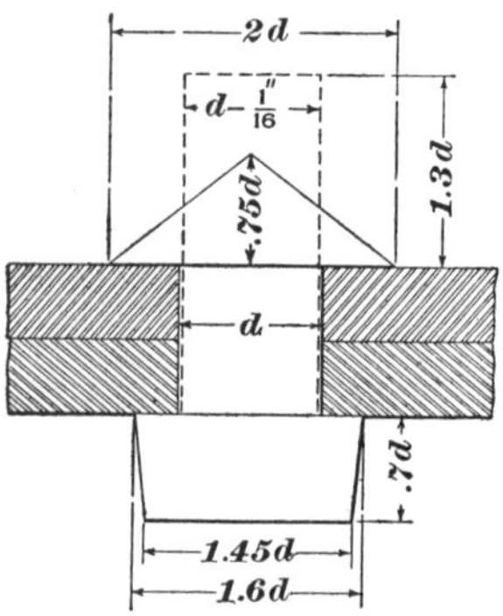

Fig. 27

The point heads shown in Figs. 27 and 28 are made by hand; this form of head is termed a **snap** or **cup head,** and is produced by hammering the shank down roughly and finishing the head with a cup-shaped die, or set. The rivets shown in Figs. 29 and 30 are examples of machine riveting. A rivet with a countersunk head is shown in Fig. 31;

such heads are sometimes necessary where a smooth surface is needed, as, for instance, when attaching boiler mountings.

Sometimes, especially on iron ships, the edges of the rivet hole are slightly chamfered, as shown in Figs. 30

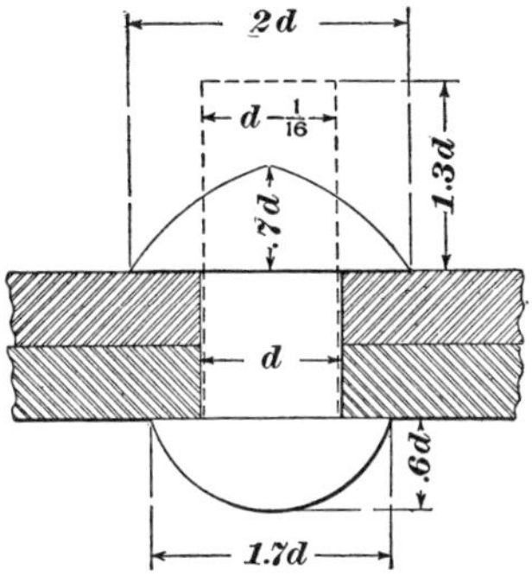

FIG. 28

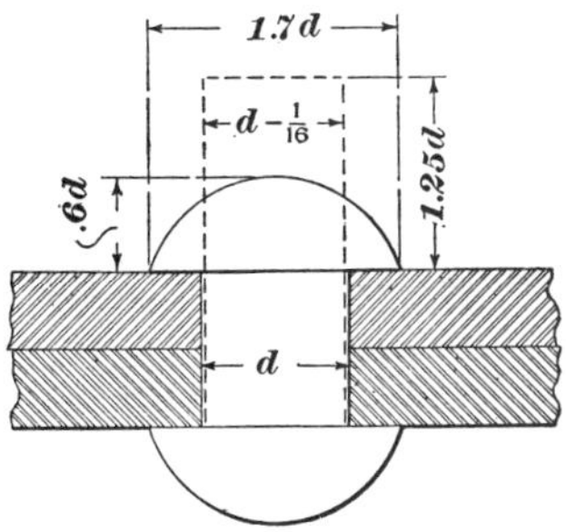

FIG. 29

and 31. A rivet driven in a hole of this kind assumes a corresponding form, resulting in fillets near the heads. Thus, the change in diameter from shank to head is made

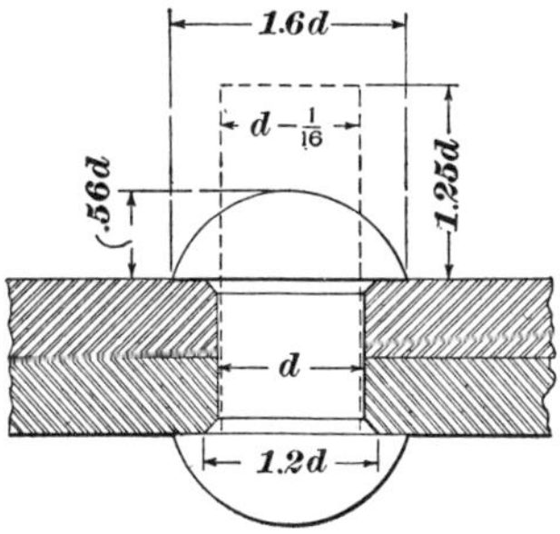

FIG. 30

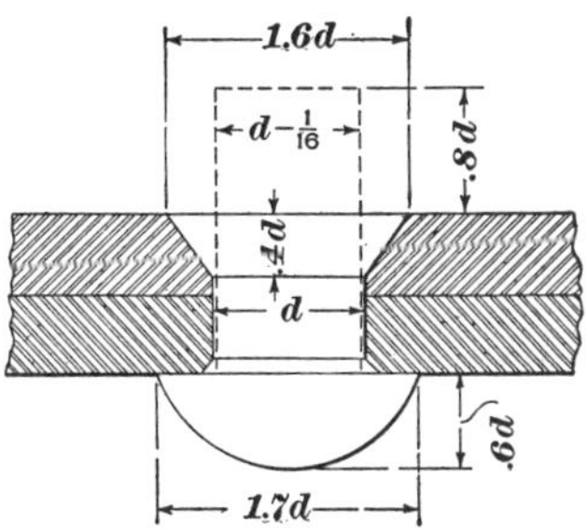

FIG. 31

more gradual, and the head is less liable to shear off than it would if subjected to the cutting action of the sharp and ragged edge of a drilled or punched hole.

CLASSIFICATION OF RIVETED JOINTS

25. Lap Joints.—The simplest form of joint is one in which the plates to be riveted are made to overlap. A joint

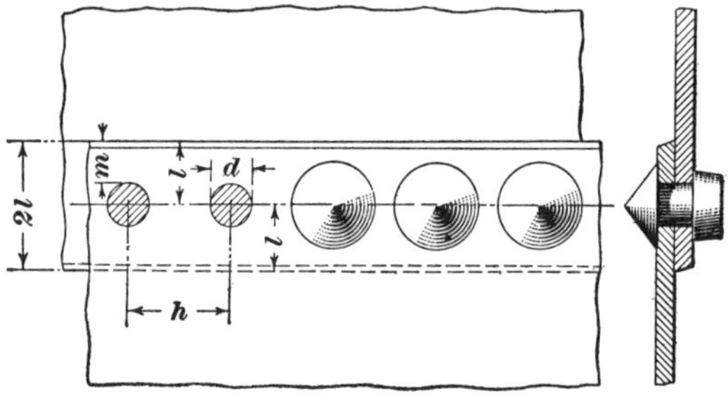

FIG. 32

of this kind is shown in Fig. 32, and is known as a **lap joint.** When a joint is made with only one row of rivets, as in Fig. 32, it is called a *single-riveted lap joint;* with two rows, as in Figs. 33 and 34, a *double-riveted lap joint;* with three rows, a *triple-riveted lap joint;* and with four rows, a *quadruple-riveted lap joint.*

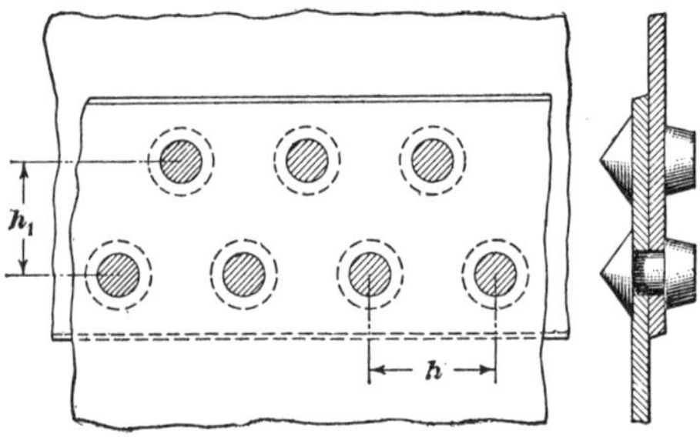

FIG. 33

When more than one row of rivets is used, distinction is made in regard to the relative arrangement of rivets in the adjoining rows. For instance, if all the centers of adjoining rivets in different rows lie in lines perpendicular to the joint, as in Fig. 34, it is called **chain riveting;** but if the rivets in one

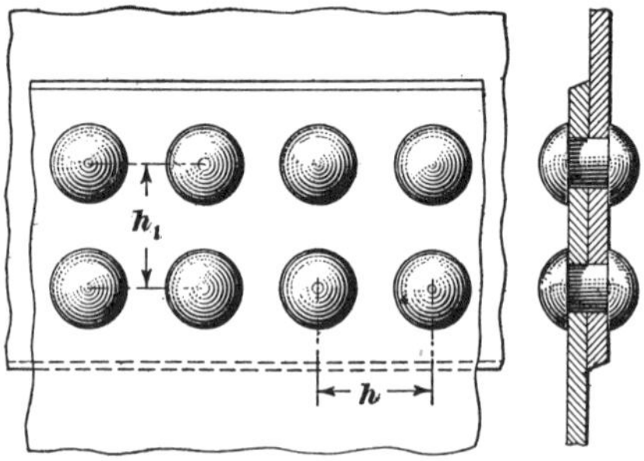

FIG. 34

row are opposite the spaces in the adjoining row, as in Fig. 33, it is called **staggered,** or **zigzag, riveting.**

A single-riveted lap joint may be provided with a *cover-plate*, or *strap*, as at *b*, *b'*, Fig. 35. Fig. 36 shows a double-riveted lap joint with cover-plate. Both joints are used extensively in boiler work, and if they are compared with the joints shown in Figs. 32 and 33, it will be seen that they are based on the latter joints, with the addition to each joint of a cover-plate and two rows of rivets spaced farther apart.

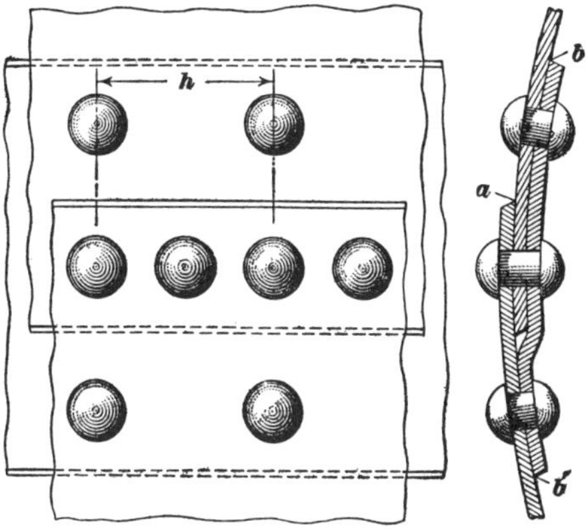

FIG. 35

26. Butt Joints.—In **butt joints,** the ends of the plates abut against each other. To effect a union, one or

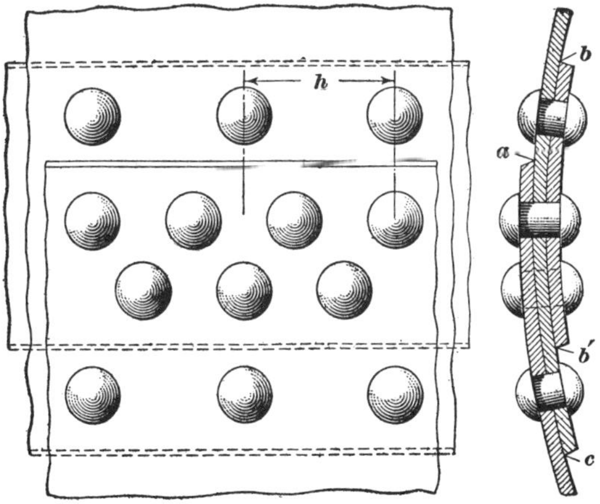

FIG. 36

more cover-plates, or straps, are required. With one cover-plate, as in Fig. 37, it constitutes a *single-strap butt joint*.

and with two cover-plates, as in Fig. 38, a *double-strap butt joint.* Butt joints are *single-riveted* when there is only one row on *each side* of the joint; with two, three, and four rows on each side, they are called *double-*, *triple-*, and *quadruple-riveted butt joints*, respectively.

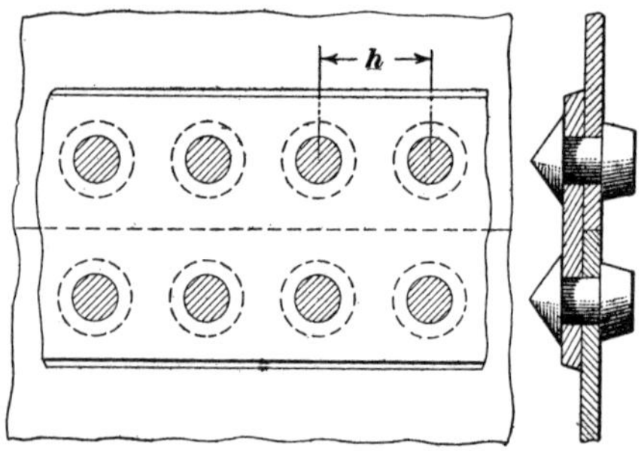

FIG. 37

Fig. 37 shows a single-riveted, and Fig. 38 a double-riveted, butt joint. The two cover-plates may differ in width, and the rivets on the outer rows may be spaced farther apart, as shown in Fig. 39.

A triple-riveted double-strap butt joint is shown in Fig. 40. The rivets through both straps are staggered, as in the double-riveted lap joint, Fig. 33.

Fig. 41 shows a quadruple-riveted double-strap butt joint

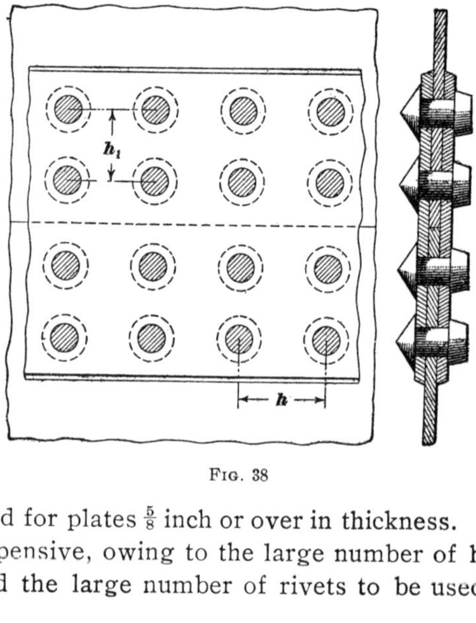

FIG. 38

that is used for plates $\frac{5}{8}$ inch or over in thickness. This joint is very expensive, owing to the large number of holes to be drilled and the large number of rivets to be used; but it is

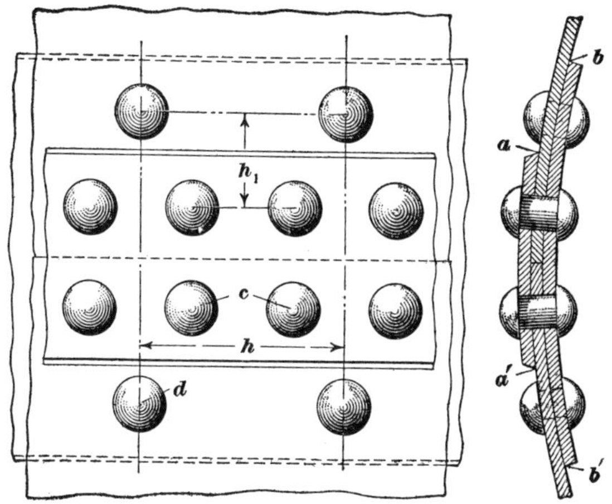

FIG. 39

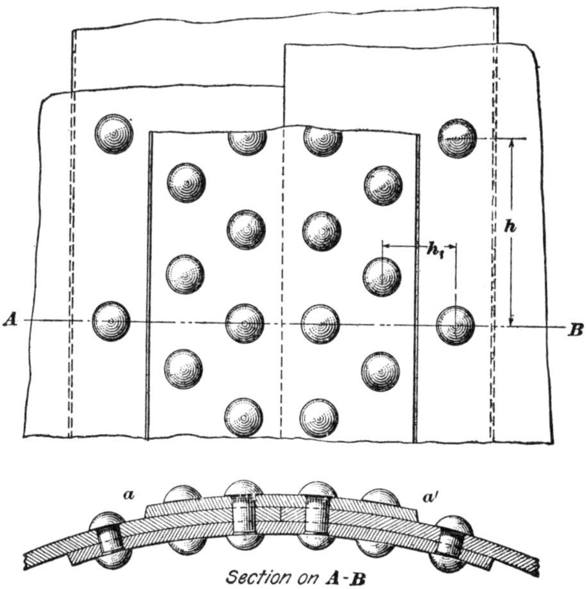

FIG. 40

one of the strongest joints to be found in practice and is also of the highest efficiency.

27. Pitch, Margin, and Lap.—The distance *h*, Fig. 32, between the centers of adjacent rivets in one row is called the **pitch** of the rivets. This pitch may vary in different

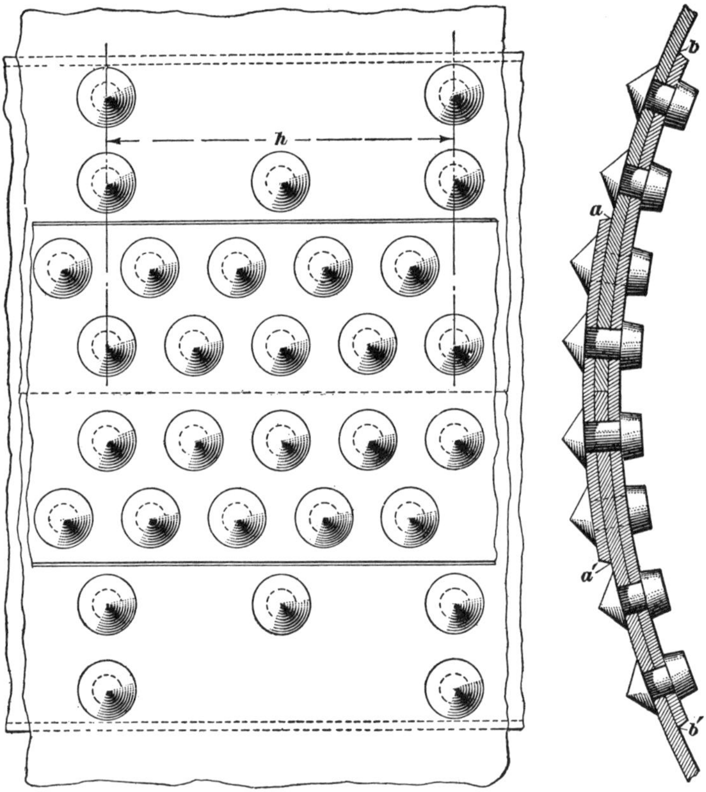

FIG. 41

rows, as shown in Fig. 35, in which the greatest distance, as shown at *h*, is considered as the pitch.

The distance *m*, Fig. 32, between the edge of the rivet hole and the edge of the plate is the **margin**, and the length $2l$ is the **lap**. The distance *l* is usually made one and one-half times the diameter *d* of the rivet hole. Recent experiments, however, have shown that when l is made only $1\frac{1}{2}d$ the margin may tear before the joint will fail in any other way. For this reason, some engineers prefer to make l equal to $2d$. The distance h_1 between two rows of rivets is called the **transverse pitch.**

28. Calking.—After the shank of the rivet has contracted, the high pressure produced on the parts between the heads is not always sufficient to insure a steam or water-tight joint where this is required. An additional treatment, termed **calking,** is then necessary. This treatment consists in forming a burr along the edge of one of the plates by means of a suitable calking tool, as shown in Fig. 42. A round-nose calking tool is driven against the beveled edge of the upper plate, forcing the metal in close contact with the lower plate and effectually closing the seam. A tool with a sharp edge should never be used, as it is liable to score the under plate and cause grooving. Such grooves may act as starting points for cracks or bends.

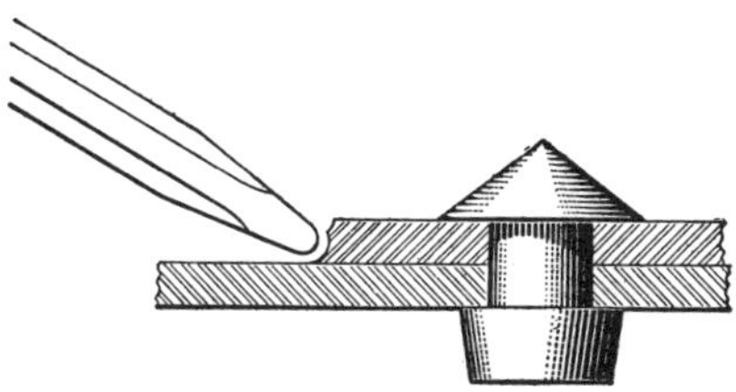

FIG. 42

On boiler joints, the calking may be done either on the inside or on the outside, or on both, the most effective calking being done on the outside. The calking edge of a plate should be beveled to make an angle of 15° with a line normal to the contact surface of the plate. To insure a good calking edge, the lap of the joint should not be excessive, as the plate will buckle instead of having its edge upset. Likewise, the pitch of the rivets should be such as to make it possible to produce a tight joint by calking.

In some of the preceding examples of riveted joints, the calking edges are indicated by the letter a. For instance, in Fig. 35, the joint is calked along the edge a of the outside plate. If the calking were done along the cover-plate, it would not only be necessary to calk two edges b, b', but the difficulty in securing a tight joint would be greater—in some cases too great—on account of the wide spacing of the adjoining rivets.

In the joint shown in Fig. 36, the calking is done at the edge *a* of the outer plate, as it would be useless to calk the cover-plate at *b*, *b′*, while at *c* the rivets are spaced too far apart to insure a tight joint.

The butt joints in Figs. 39, 40, and 41 are calked at the edges *a*, *a′*, of the outside cover-plates, as the other cover-plates are inside and have rivets spaced too far apart.

STRENGTH OF RIVETED JOINTS

29. Failure of Riveted Joints.—When the load applied to the rivet joint acts in a direction at right angles to the axes of the rivets, the joint may fail in one of the following five ways, which are illustrated in Fig. 43.

1. The plate may fracture along the center line of the rivets, as shown in (*a*).

2. The rivet may shear, as shown in (*b*).

3. The rivet or the plate in front of it may be crushed, as shown in (*c*).

4. The plate may shear in front of the rivets, as shown in (*d*).

5. The plate may break across the margin, as shown in (*e*).

To ascertain which part of a rivet joint is the weakest and where, therefore, a failure is likely to occur, each part of the joint will be considered separately and a formula representing the strength at this particular point will be given.

Let d = diameter of rivet *after* riveting, or diameter of rivet hole;

h = pitch of rivets, in inches; where the pitch varies, the largest pitch is taken;

t = thickness of plate, in inches;

f = factor of safety;

n = number of rivets for a length of joint equal to h;

l = distance from center of rivet to edge of plate, in inches, usually taken as $1\frac{1}{2}\,d$;

S_t = ultimate tensile strength, in pounds per square inch of plate = 55,000 pounds for steel and 50,000 pounds for wrought iron;

S_c = ultimate crushing strength per square inch of the plate = 90,000 pounds for steel and 80,000 pounds for wrought iron;

S_s = ultimate shearing strength per square inch of rivets = 45,000 pounds for steel and 40,000 pounds for wrought iron;

R_t, R_c, and R_s = ultimate strength of riveted joint for a width equal to h, for tension, compression, and shearing, respectively.

The values of the ultimate strengths just given are average values for materials used in riveted joints. Whenever possible these values should be replaced by values found by actual tests of the material to be used.

30. Strength of Lap Joints.—Beginning with the first case, illustrated in Fig. 43 (*a*), the possible modes of failure mentioned in Art. **29** will be considered in detail. By referring to Fig. 32, it will be seen that the area of metal between two rivet holes is $(h - d)t$; its resistance to rupture is $(h - d)\,t\,S_t$; therefore,

$$R_t = (h - d)\,t\,S_t \qquad (1)$$

In the second case, Fig. 43 (*b*), the resistance of the rivets to shear is the next point to be considered as a possible cause of failure. The shearing area of a rivet is $.7854\,d^2$, and as there is only one rivet in the length h, the shearing resistance of the rivets is equal to $.7854\,d^2\,S_s$, or,

$$R_s = .7854\,d^2\,S_s \qquad (2)$$

In the third case, Fig. 43 (*c*), the resistance to crushing of the parts located in front of each rivet will depend on the *projected crushing area* of the rivet; that is, the area of the projection of the rivet shank on a plane parallel to its axis. In the present case, this area is $d\,t$. The resistance to crushing for the length h is $d\,t\,S_c$, and

$$R_c = d\,t\,S_c \qquad (3)$$

In the fourth case, Fig. 43 (d), the resistance to shear of those parts of the plate situated in front of the rivet holes is considered. Here, the distance l, Fig. 32, between the center of the rivet hole and the edge of the plate must be taken into consideration. The part exposed to shear is nearly of prismatic form, having a width nearly equal to d, with a length l. As shearing will take place along two sides

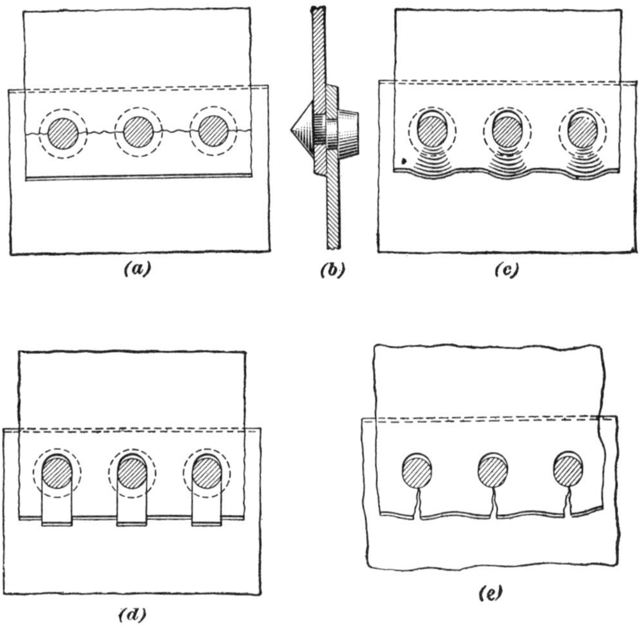

FIG. 43

of this prism, the total shearing area is $2\,l\,t$, and the resistance to shear is

$$R_s = 2\,l\,t\,S_s \qquad (4)$$

The requirement necessary to prevent the fracture considered in the fifth case, Fig. 43 (e), will depend somewhat on the nature of the material and whether the holes were punched or drilled. Considering the part in front of the rivet as a beam subjected to bending, calculations show that the margin m, Fig. 32, should be equal to $.4\,d$, in order that its resistance to fracture will equal the shearing strength of the rivet. As it is common practice to make m at least equal to d, the resistance to fracture at this place is reasonably ample.

31. If the margin of a joint is ample, it is customary to consider only the first, second, and third cases, represented by formulas **1, 2,** and **3,** Art. **30.** In order that each part of the joint will be equal in strength, the values of R, as found by these formulas, should be equal. The suitable values of d, t, and h may then be found by equating the values of R_t, R_s, and R_c.

32. In determining the diameter of the rivets, it is desirable to find one that will give the rivet equal resistance against shearing and crushing. This may be done by equating R_s and R_c in formulas **2** and **3,** Art. **30.** Thus,

$$.7854\,d^2 S_s = d\,t\,S_c$$

and

$$d = \frac{t\,S_c}{.7854\,S_s} = 1.27\,\frac{t\,S_c}{S_s} \qquad (1)$$

As the shearing resistance of a rivet varies as the square of its diameter, and the resistance to crushing varies directly as the diameter, it follows that if the diameter is made smaller than that found by formula **1,** the resistance to crushing will exceed that of shearing, and if the diameter is greater, its shearing resistance will be the greater.

To find the distance h that will give the plate a tensile strength between rivet holes equal to the shearing resistance of the rivets, combine formulas **1** and **2,** Art. **30;** or,

$$(h - d)\,t\,S_t = .7854\,d^2 S_s$$

Then,

$$h = .7854\,d^2\,\frac{S_s}{t\,S_t} + d \qquad (2)$$

In this instance, it is assumed that the diameter of the rivet is equal to or smaller than that found by formula **1,** and that, therefore, its resistance to shearing equals or is smaller than that to crushing. If the rivet has a diameter *larger* than that found by formula **1,** its crushing resistance must be taken into account, and the values of R_t and R_c in formulas **1** and **3,** Art. **30,** must be equated. Then,

$$(h - d)\,t\,S_t = d\,t\,S_c$$

and

$$h = \frac{d(S_t + S_c)}{S_t} \qquad (3)$$

33. In double-riveted joints, such as shown in Figs. 33 and 34, all rivets contained in a length h of the joint (see Fig. 34) are supposed to participate in the shearing and crushing stresses. Therefore, in Figs. 33 and 34, there will be two rivets subjected to shearing, and the crushing stress will be distributed along a length of plate corresponding to that located on front of two rivets. In determining the tensile strength of the plate, only that part included between two rivets in one row is considered, and formula **1,** Art. **30,** may therefore be used for this joint also.

The shearing resistance of the rivets is

$$R_s = 2 \times .7854\, d^2\, S_s = 1.57\, d^2 S_s \qquad (1)$$

The resistance of the joint to crushing is

$$R_c = 2\, d\, t\, S_c \qquad (2)$$

If the plates have the same thickness as those of a single-riveted lap joint, the rivets may either be placed farther apart or they can be of smaller diameter.

Equating the values of R_s and R_c,

$$d = 1.27\, \frac{t\, S_c}{S_s} \qquad (3)$$

Equating the values of R_t and R_s,

$$h = \frac{1.57\, d^2 S_s}{t\, S_t} + d \qquad (4)$$

If d is taken *larger* than the value found by formula **3,** then, following the method given in Art. **32,**

$$h = \frac{d(2\, S_c + S_t)}{S_t} \qquad (5)$$

The transverse pitch h_1, Fig. 33, should have a minimum value of $.4\, h$; it is generally $.5\, h$ to $.65\, h$. For the joint, Fig. 34, h_1 should not be less than $2\, d$ and is preferably $2.5\, d$ for boilers and $3\, d$ for structural work.

EXAMPLE 1.—A single-riveted joint is to be made with $\frac{5}{16}$-inch wrought-iron plates. Find the rivet pitch and the diameter of the rivet hole.

SOLUTION.—From formula **1,** Art. **32,** $d = 1.27\, \frac{t\, S_c}{S_s}$. From Art. **29,** S_s = 40,000 lb., and S_c = 80,000 lb.; hence,

$$d = 1.27 \times \tfrac{5}{16} \times \frac{80{,}000}{40{,}000} = .79 \text{ in.}$$

If the nearest smaller size available is $\frac{3}{4}$ in., then formula **2,** Art. **32,** is used for ascertaining h; if the next larger size, such as $\frac{13}{16}$ in., then formula **3,** Art. **32,** is used. As $d = \frac{3}{4}$ in.,

$$h = .7854 \times (\tfrac{3}{4})^2 \times \frac{40{,}000}{\frac{5}{16} \times 50{,}000} + \tfrac{3}{4} = 1.88, \text{ or } 1\tfrac{7}{8}, \text{ in., nearly}$$

If $d = \frac{13}{16}$ in.,

$$h = \frac{\frac{13}{16} \times (50{,}000 + 80{,}000)}{50{,}000} = 2.113, \text{ or } 2\tfrac{1}{8}, \text{ in., nearly. Ans.}$$

EXAMPLE 2.—If a double-riveted lap joint is to be made with the plates in example 1, what will be the pitch?

SOLUTION.—From formula **3,** Art. **33,** it is seen that the rivet diameter remains the same. From formula **4,** Art. **33,**

$$h = \frac{1.57 \times (\frac{3}{4})^2 \times 40{,}000}{\frac{5}{16} \times 50{,}000} + \tfrac{3}{4} = 3 \text{ in. Ans.}$$

34. Strength of Butt Joints.—The butt joint with a single cover-plate, shown in Fig. 37, consists practically of two lap joints, each being a joint between the strap and a plate. The formulas applying to a single lap joint may therefore be used for each half of the butt joint. To prevent the cover-plate from bending, its thickness is increased by at least one-eighth that of the plates, and its tensile and crushing resistances will be greater.

The butt joint with two cover-plates, shown in Fig. 38, has an advantage over a lap joint or over a butt joint with one cover-plate, in that it is free from the tendency to bend the plates. In addition to this, the rivets are subjected to double shear. This style of joint is therefore considered very strong, but it is expensive on account of the work required to drill so many holes and to drive so many rivets. In first-class work, however, the required strength of joint, and not the cost, should determine the type of joint to be used.

The tensile resistance of the plate along the outer row of rivets is the same as that of a lap joint, and is found by means of formula **1,** Art. **30,**

$$R_t = (h - d)\, t\, S_t \qquad \textbf{(1)}$$

There are two rivets in a length h subjected to double shear. The fact that these rivets are in double shear does not make them twice as strong as rivets in single shear. The ratio between the strength of a rivet in double shear to that in single shear is 1.75, according to the British Board of Trade and the Canadian rules, while the Hartford Steam Boiler Insurance and Inspection Company uses and advocates a ratio of 1.85. In the succeeding calculations, a value of 1.75 will be used. The total resistance of the rivets is therefore taken as

$$R_s = 2 \times 1.75 \times .7854\, d^2 S_s = 2.75\, d^2 S_s \qquad (2)$$

The resistance of the plate to crushing will be

$$R_c = 2\, d\, t\, S_c \qquad (3)$$

The value of d that gives the rivet an equal shearing and crushing resistance is found by equating formulas **2** and **3**, or

$$2.75\, d^2 S_s = 2\, d\, t\, S_c$$

when

$$d = .73\, t \frac{S_c}{S_s} \qquad (4)$$

To ascertain the pitch that will give the joint equal strength against rupture and crushing, formulas **1** and **3** may be equated as follows:

$$(h - d)\, t\, S_t = 2\, d\, t\, S_c$$

when

$$h = \frac{2\, d\, S_c}{S_t} + d \qquad (5)$$

The combined thickness of the butt straps should be more than that of the plate. Single butt straps should be at least $\frac{9}{8}\, t$. For double straps, the minimum thickness should be $\frac{5}{8}\, t$, if no rivets are omitted in the outer row. In the latter case, the straps should be increased in thickness in the ratio $\frac{h - d}{h - 2\, d}$. For the joint in Fig. 41, calculations show that the straps should be $\frac{3}{4}\, t$. When the joint must be calked, the outer strap should not be made less than from $\frac{1}{4}$ to $\frac{5}{16}$ inch thick, and both straps are usually made of the same thickness. If the straps are of unequal thickness, it is customary to calculate the thickness of the thinner strap as if both straps were of the same thickness.

35. The double-riveted butt joint shown in Fig. 39 differs from the joint shown in Fig. 38 in that the rivets in the outer rows do not pass through the outside cover-plate, and that they have a greater pitch than those in the inner rows. In this instance, the distance h between the rivets in the outer rows is generally taken as the pitch and as the width of the unit strip.

The joint may fail by tearing along the outer rows of rivets, and the strength of the plate required to resist this failure may therefore be determined by formula **1**, Art. **30**, that is,

$$R_t = (h - d)\,t\,S_t \qquad (1)$$

The failure might occur by shearing two rivets c, Fig. 39, in double shear and one rivet d in single shear. As in Fig. 38, the resistance of a rivet in double shear is taken as 1.75 times its resistance in single shear. The resistance of the joint to shear will therefore be that of two rivets in double shear, equal to $2 \times 1.75 = 3.5$ in single shear, plus one rivet in single shear, or of 4.5 rivets in single shear. Then, the resistance to shear is

$$R_s = 4.5 \times .7854\,d^2 S_s = 3.53\,d^2 S_s \qquad (2)$$

The failure might be caused by tearing the plate along the inner row of rivets and shearing the outer row of rivets in single shear. Then, if R_{ts} represents the resistance to tension and shearing combined,

$$R_{ts} = (h - 2\,d)\,t\,S_t + .7854\,d^2 S_s \qquad (3)$$

The failure might be from crushing the plate in front of the three rivets in the distance h. Then,

$$R_c = 3\,d\,t\,S_c \qquad (4)$$

Finally, the failure might be from crushing the plate in front of two inner rivets and shearing one of the outer rivets in single shear. Then, letting R_{cs} represent the resistance to crushing and shearing combined,

$$R_{cs} = 2\,d\,t\,S_c + .7854\,d^2 S_s \qquad (5)$$

Equating the values of R_s and R_c, then

$$3.53\,d^2 S_s = 3\,d\,t\,S_c$$

and

$$d = .85\,t\,\frac{S_c}{S_s} \qquad (6)$$

Equating the values of R_t and R_s, then

$$(h - d)\,t\,S_t = 3.53\,d^2 S_s$$

and

$$h = \frac{3.53\,d^2 S_s}{t\,S_t} + d \qquad (7)$$

For double-riveted butt joints with zigzag riveting, and for chain riveting with alternate rivets omitted in the outer row, the transverse pitch is usually

$$h_1 = \frac{4d + 1}{2}$$

For zigzag riveting with alternate rivets omitted in the outer row, similar to Fig. 39, the least value of

$$h_1 = \frac{\sqrt{(11h + 20d)(h + 20d)}}{20} \qquad (8)$$

36. Efficiency of a Riveted Joint.—By **efficiency** of a riveted joint is meant the ratio of its strength to that of the solid plate. In the preceding calculations of strength of joints, the pitch and the dimensions of rivets and plates were chosen so as to make the joint of equal strength against tensile, shearing, and crushing stresses. It is then immaterial which one of these strengths is compared with the strength of the solid plate in order to ascertain the efficiency, though it is customary to consider the strength of the plate between the two rivets in the outer row. Considering a width h, Fig. 32, the area of the solid plate is $h\,t$, and that of the metal between the rivet holes is $(h - d)\,t$. If y is the efficiency of the joint, then

$$y = \frac{(h - d)\,t}{h\,t}$$

or

$$y = \frac{h - d}{h}$$

The efficiency of the riveted joint should be considered for all cases of riveting.

EXAMPLE.—In a single-riveted lap joint, the rivet holes are $\frac{11}{16}$ inch in diameter and the pitch of the rivets is 2 inches. What is the efficiency of the joint?

SOLUTION.—Applying the formula $y = \frac{h - d}{h}$,

$$y = \frac{2 - \frac{11}{16}}{2} = .66, \text{ nearly. Ans.}$$

37. Practical considerations often demand that some of the dimensions be changed; consequently, the three values of R will no longer be equal. For instance, a riveted joint of sufficient strength may have a distance between rivets so great that it will be difficult to secure a tight joint. Smaller rivets, more closely spaced, may then be required. Likewise, it is possible that the commercial sizes of plates and rivets may not correspond to those demanded by the formulas, and that other sizes must be substituted. In such cases, the formula $y = \frac{h - d}{h}$ would not represent the efficiency of the joint, and it would be necessary to ascertain also the other efficiencies. The lowest one of these efficiencies would then be considered as that of the joint.

By increasing the number of rivet rows in combination with an increased pitch in the outer row, the efficiency of a joint is generally increased. For this reason, Fig. 41 represents one of the most efficient joints in practical use, but it is also an expensive one on account of the number of holes and rivets required.

38. Strength of Cylindrical Shells.—Before showing the application of the preceding formulas, it will be necessary to state briefly the requirements as to the thickness of shells exposed to internal pressure, such as boilers, reservoirs, etc. The ultimate strength of such shells may be computed by using the formula $pD = 2tS_t$, the derivation of which was shown in *Strength of Materials*, Part 2. In this formula,

P = pressure, in pounds per square inch;
D = diameter of shell, in inches;
t = thickness of shell, in inches;
S_t = ultimate tensile strength, in pounds per square inch;
f = factor of safety.

Then, the safe working stress is $\frac{S_t}{f}$ and

$$pD = \frac{2tS_t}{f}$$

For shells subjected to internal pressure, the factor f is generally taken from 4 to 6.

Solving for S_t in the formula just given,

$$S_t = \frac{p\,D\,f}{2\,t} \qquad \textbf{(1)}$$

Solving for t,

$$t = \frac{p\,D\,f}{2\,S_t} \qquad \textbf{(2)}$$

In the riveted joints used on a shell, distinction is made between those that run parallel with the axis and others that run around the shell. The first are termed **longitudinal seams** and the latter **girth seams.**

The length of a girth seam is equal to the circumference of the shell, or 3.1416 D, and the sectional area of the shell along a girth seam is 3.1416 $D\,t$. Its ultimate resistance to rupture is 3.1416 $D\,t\,S_t$, and its safe working strength 3.1416 $D\,t\frac{S_t}{f}$.

The area on which the pressure p acts to produce a girth rupture is .7854 D^2; hence, the pressure tending to produce a rupture is .7854 $D^2\,p$. For the area located along the line where rupture would take place, the two expressions indicating pressure and resistance must be equal. Therefore,

$$.7854\,D^2 p = 3.1416\,D\,t\frac{S_t}{f}$$

or

$$S_t = \frac{p\,D\,f}{4\,t} \qquad \textbf{(3)}$$

Formulas **1** and **3** prove that for a given pressure the stress in the shell is twice as great in the longitudinal seam as it is in the girth seam. For this reason, the efficiency of a girth seam need be only one-half as great as the efficiency of the longitudinal seam, in order to furnish equal resisting strength to a given internal pressure; this explains why girth seams are single-riveted, while longitudinal seams of the same shell are double-riveted.

39. Design of Riveted Joints.—As an example of the application of the formulas relating to riveted joints, let it be required to design the riveting and to find the thickness of the shell intended for a compressed-air reservoir 3 feet 9 inches in diameter and working under a pressure of 160 pounds per square inch, assuming a factor of safety of 6. The longitudinal seam is to be a double-riveted butt joint, as shown in Fig. 39, and the girth seam, a single-riveted lap joint. The factor of safety refers to the solid plate. In a perforated plate, the factor will be smaller, depending on the efficiency of the joint. If this efficiency is about 80 per cent., the actual factor of safety will be nearer 5.

It is assumed that the material to be used is steel, and that its properties have been ascertained by special tests, whereby the following values were found: $S_t = 55{,}000$ pounds; $S_c = 90{,}000$ pounds; and $S_s = 45{,}000$ pounds.

The thickness of the shell will be found from formula **2**, Art. **38**; thus,

$$t = \frac{160 \times 45 \times 6}{2 \times 55{,}000} = .39 \text{ inch}$$

The nearest commercial size is $\frac{7}{16}$ inch, which is the value given t.

By equating the values of R_s and R_c in formulas **2** and **4**, Art. **35**, the smallest allowable value for the diameter d for this style of joint will be obtained. Thus

$$3.53\, d^2 S_s = 3\, d\, t\, S_c$$

On solving for d,

$$d = \frac{3\, t\, S_c}{3.53\, S_s} = \frac{3 \times .4375 \times 90{,}000}{3.53 \times 45{,}000} = .74 \text{ inch}$$

As this is the smallest allowable size, a $\frac{3}{4}$-inch rivet would do, but the driven size had better be taken larger; a $\frac{13}{16}$-inch rivet hole is therefore preferable.

By equating the values of R_{ts} and R_{cs} in formulas **3** and **5**, Art. **35**, a value for h can be obtained. Thus,

$$(h - 2\, d)\, t\, S_t + .7854\, d^2 S_s = 2\, d\, t\, S_c + .7854\, d^2 S_s$$

Then,
$$(h - 2\, d)\, t\, S_t = 2\, d\, t\, S_c$$

and
$$h - 2\, d = \frac{2\, d\, S_c}{S_t}$$

from which

$$h = 2\, d\left(1 + \frac{S_c}{S_t}\right) = 1\tfrac{5}{8} \times \left(1 + \frac{90{,}000}{55{,}000}\right)$$
$$= 4.284 = 4\tfrac{5}{16} \text{ in., nearly}$$

From formula **8**, Art. **35**,

$$h_1 = \frac{\sqrt{(11 \times 4\tfrac{5}{16} + 20 \times \tfrac{13}{16}) \times (4\tfrac{5}{16} + 20 \times \tfrac{13}{16})}}{20}$$

$$= 1.81, \text{ or } 1\tfrac{13}{16}, \text{ inches, nearly}$$

40. From the formulas given in Art. **35**, the values R_t, R_s, R_{st}, R_c, and R_{cs} may now be calculated. The smallest value will determine the efficiency of the joint.

From formula **1**, Art. **35**,

$$R_t = (4\tfrac{5}{16} - \tfrac{13}{16}) \times \tfrac{7}{16} \times 55{,}000 = 84{,}219 \text{ pounds}$$

From formula **2**, Art. **35**,

$$R_s = 3.53 \times (\tfrac{13}{16})^2 \times 45{,}000 = 104{,}860 \text{ pounds}$$

From formula **3**, Art. **35**,

$$R_{st} = (4\tfrac{5}{16} - 1\tfrac{5}{8}) \times \tfrac{7}{16} \times 55{,}000 + .7854 \times (\tfrac{13}{16})^2 \times 45{,}000$$
$$= 88{,}000 \text{ pounds}$$

From formula **4**, Art. **35**,

$$R_c = 3 \times \tfrac{13}{16} \times \tfrac{7}{16} \times 90{,}000 = 95{,}977 \text{ pounds}$$

From formula **5**, Art. **35**,

$$R_{cs} = 2 \times \tfrac{13}{16} \times \tfrac{7}{16} \times 90{,}000 + .7854 \times (\tfrac{13}{16})^2 \times 45{,}000$$
$$= 87{,}316 \text{ pounds}$$

Of these values, R_t is the smallest, and will therefore determine the efficiency of the joint. The strength of the solid plate is

$$h\,t\,S_t = 4\tfrac{5}{16} \times \tfrac{7}{16} \times 55{,}000 = 103{,}770 \text{ pounds}$$

Hence, the efficiency of the joint is

$$y = \frac{84{,}219}{103{,}770} = .81$$

From Art. **38**, the efficiency of the girth seam need not be more than one-half that of the longitudinal seam. Retaining rivets of the same size, and making the pitch h in the single-riveted lap joint one-half that of the longitudinal seam,

$$h = \frac{4\tfrac{5}{16}}{2} = 2\tfrac{5}{32} \text{ inch}$$

The efficiency is

$$y = \frac{2\tfrac{5}{32} - \tfrac{13}{16}}{2\tfrac{5}{32}} = .62$$

As this is considerably more than half the efficiency of the longitudinal seam, the girth seam will be amply strong.

The real factor of safety of the shell is affected by the efficiency of the riveted joint. The safe strength of the shell at the longitudinal joint for a width h is $(h - d)\,t\frac{S_t}{f}$, and for the two sides of the shell $2(h - d)\,t\frac{S_t}{f}$. This value must be equal to the internal pressure for the length h, which is $h p D$. Hence,

$$h p D = 2(h - d)\,t\frac{S_t}{f}$$

and

$$f = \frac{2(h - d)\,t\,S_t}{h p D} = \frac{2(4\tfrac{5}{16} - \tfrac{13}{16}) \times \tfrac{7}{16} \times 55{,}000}{4\tfrac{5}{16} \times 160 \times 45} = 5.42$$

If the joint is perfect in design and workmanship, the strengths at the various possible places of failure should be equal; or, to be on the safe side, the value R_s should be greater, as it is not always practicable to make the rivet holes in the shell plates and cover-plates come absolutely fair; consequently, the rivet sections would be reduced. As a rule, perfect equality of the strengths is difficult to obtain, because the commercial sizes of plates and rivets must be used; and these do not always correspond to the sizes required by the formulas.

41. In the joint under consideration, it is assumed that the cover-plates are of such thickness as to prevent their failure before the steel plate. To insure this condition, the thickness of each cover-plate should generally be at least five-eighths that of the shell. In this instance, therefore, they should be at least $\frac{7}{16} \times \frac{5}{8} = \frac{35}{128}$ inch. The nearest value to this in sixteenths is $\frac{5}{16}$ inch.

The mode of failure of the cover most likely to occur is the tearing of the inside cover-plate at the inner row of rivets, the shell plate freeing itself by shearing the rivets in the outside cover-plate.

The resistance of the cover-plate is

$$(h - 2d)\,t\,S_t = (4\tfrac{5}{16} - 1\tfrac{5}{8}) \times \tfrac{5}{16} \times 55{,}000 = 46{,}191 \text{ pounds}$$

The resistance of the two rivets in shear is

$$2 \times .7854\,d^2\,S_s = 2 \times .7854 \times (\tfrac{13}{16})^2 \times 45{,}000 = 46{,}664 \text{ pounds}$$

The total resistance is 46,191 + 46,664 = 92,855 pounds, which is greater than R_t.

For a quadruple-riveted joint, such as is shown **in Fig. 41**, the thickness of the cover-plate must be at least three-fourths as thick as the shell plate; otherwise, the efficiency will be lowered by the mode of failure just mentioned.

It should always be remembered that the pitch and rivet diameter found efficient for a shell plate of a given thickness and strength may be entirely unsuited for a plate having different thickness and strength.

42. Tables of Riveted Joints.—Table IV, at the end of this Section, gives the proportions of riveted joints used by a number of the leading manufacturers in the United States. The table is applicable to lap joints, and butt joints with a single cover-plate. For boiler plates more than $\frac{1}{2}$ inch thick, butt joints with two cover-plates are recommended. The efficiencies y in this table are obtained from the formula of Art. **36.** The terms *single* and *double* refer to single- and double-riveted joints, respectively. For example, the efficiency of a double-riveted joint made of $\frac{7}{16}$-inch plates is .74, and the pitch h is $3\frac{3}{8}$ inches. In calculating the efficiencies given in Table IV, it is supposed that the proportions of the joints and the properties of the material are such that the single-riveted joints will fail by tearing along the single row of rivets, and the double-riveted joints by tearing along the outer row of rivets.

Table V, at the end of this Section, gives the size, arrangement of rivets, and width of lap for various thicknesses of sheets in single-riveted lap joints, such as used on tanks, stacks, etc. For rectangular tanks, this table also gives the distance between the center of rivet and outside of tank, likewise the radius of curvature of the corners in same.

STRUCTURAL RIVETING

43. Distinctive Features.—**Structural riveting** deals with the fastening of rolled shapes, such as beams, angles, etc., to one another or to plates by means of rivets. This class of riveting has much in common with that used where internal pressure must be resisted, both in regard to the kinds of joints and the shapes of the rivets. The difference between them lies in the fact that structural riveting allows more freedom in the selection of pitch; the distance between the rivets is of secondary importance, the main object being to employ a sufficient number of rivets to gain the requisite strength. The pitch, however, should not be so great as to allow moisture to enter the joint, resulting in the production of rust, and should rarely exceed 6 inches. The minimum pitch is determined by the necessary clearance of the riveting tool, and by the danger of causing cracks in the plate between the holes during the punching or riveting process.

44. Rivet Heads.—The forms of rivets found in structural work are of less variety then those used in shells and tanks. The principal forms are: (1) the **button head,** also called **round head,** or **full head;** (2) the **countersunk head;** and (3) the **flattened head,** which is made by flattening the button head, and is used in extreme cases where a round head would be too high.

45. Shop and Field Rivets.—Distinction is made between *shop rivets* and *field rivets.* **Shop rivets** are those driven in place in the shop, and **field rivets,** those that have to be inserted at the place of erection. Shop rivets are made of soft steel, while field rivets are of wrought iron, because this material is less likely to be damaged in heating, and also because considerable heat is lost in passing the rivets from the forge to the riveters, thus reducing them to a temperature that would be too low to work on steel.

46. Diameter of Rivets.—Rivets used in structural work vary in diameter from $\frac{3}{8}$ inch to 1 inch, being made in

$\frac{3}{8}$-, $\frac{1}{2}$-, $\frac{5}{8}$-, $\frac{3}{4}$-, $\frac{7}{8}$-, and 1-inch sizes. The $\frac{3}{4}$- and $\frac{7}{8}$-inch rivets are most frequently used, the smaller sizes being employed only on unimportant details, or where the clearance will not allow the insertion of larger ones. Rivets larger than $\frac{7}{8}$ inch are used only where the plates are very thick or the stresses very great. Field rivets should never be more than $\frac{7}{8}$ inch.

On account of the difficulty in punching holes of smaller diameter than the thickness of the plate, the diameter of the rivet should not be less than the thickness of the thickest plate through which it passes. Rolled sections with small flanges necessitate the use of small rivets, so as to allow enough margin and sufficient space for the tool in driving the rivet. When the size is not determined by the conditions just mentioned, a rivet should be chosen that will resist the existing stresses and at the same time be the least expensive. It is cheaper to use one kind of rivet throughout a piece, and the smaller rivets are less expensive, as they reduce the weight and the cost of punching and driving. The standard sizes used by the Cambria Steel Company in **I** beams, channels, and angles are given in Table VI, at the end of this Section.

As to the relations between the diameters of rivets and the corresponding rivet holes, the explanations given in Art. **20** hold good. Holes are not drilled except in very important work, or when hard steel is used.

47. Dimensions of Heads.—The height and diameter of a rivet head bear the following relations to the diameter of the rivet:

For button heads, when d is the diameter of the rivet, the standard height of the head is $.6\,d$, and the radius of the head is $\frac{3}{4}d + \frac{1}{16}$ inch. These dimensions are shown in Fig. 44.

For countersunk heads, the diameter of the head is the same as for a button head, and its bevel is 60°. A simple rule is to make the depth equal to $\frac{1}{2}d$. Countersunk rivets should not be used in plates having a thickness less than $\frac{1}{2}d$. Where economy is to be considered, their use should be avoided if possible, as the extra work in countersinking makes them expensive.

For flattened heads, the height is never less than $\frac{5}{16}$ inch.

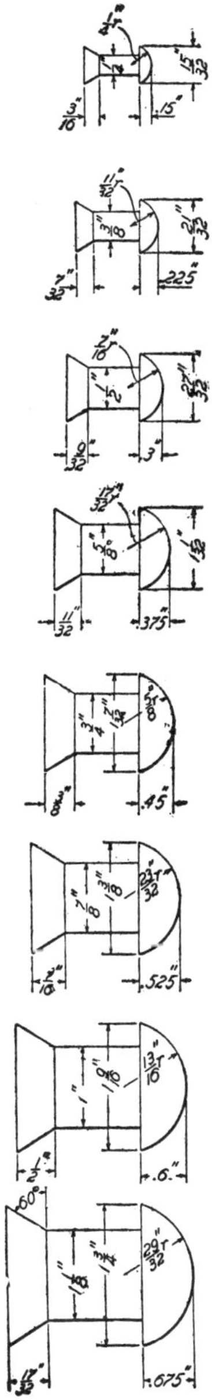

FIG. 44

48. Length, or Grip.—The maximum **grip,** or distance between heads, is generally accepted as not more than four times the diameter of the rivet. In determining the length to be given in ordering rivets, first find the grip, which is considered as the thickness of the parts joined plus $\frac{1}{32}$ inch for each joint between plates; this $\frac{1}{32}$ inch is to allow for unevenness in the surfaces of the plates, which might prevent them from being in close contact. This grip must be increased in the ratio of the area of the hole to the area of the rivet section, to allow for upsetting, the hole being $\frac{1}{16}$ inch larger in diameter than the rivet. To this result must be added the height required for the head, and the sum is the length of rivet under the formed head.

The values given in Table VII, at the end of this Section, have been calculated in the manner just described. To illustrate the application of this table, two examples are given.

It is desired to find the shank length of a $\frac{3}{4}$-inch button-head rivet with a grip of $2\frac{1}{8}$ inches. Proceed horizontally from the value $2\frac{1}{8}$ in the first column of the table to the vertical column headed $\frac{3}{4}$; in this column the value $3\frac{5}{8}$ is found, which is the required length of the shank.

If the shank length of a countersunk rivet is to be ascertained, a certain amount has to be deducted from the value given in the table. The amount of this deduction is found at the foot of the table. For instance, if the shank mentioned in the preceding example is that of a countersunk rivet, then the amount $\frac{5}{8}$, which is found at the foot of the vertical column headed $\frac{3}{4}$, is to be deducted. The shank length of a $\frac{3}{4}$-inch countersunk rivet is therefore $3\frac{5}{8} - \frac{5}{8} = 3$ inches.

NOTE.—The right-hand illustration at the head of Table VII indicates the length that is considered as the grip of a countersunk rivet.

49. Pitch.—The **pitch,** or distance from center to center of rivets, should be at least three diameters of the rivet. If spaced closer, the material is likely to fracture or become otherwise injured. For members in compression, the pitch in the direction of the stress should not be greater than 6 inches, or sixteen times the thickness of the thinnest outside plate. If possible, the pitch should be kept in even inches, making it either 2, 3, 4, 5, or 6 inches, as the case may require, especially when the rows are long.

In Table VIII, at the end of this Section, are given the maximum and minimum pitches for rivets of different sizes and the minimum and usual distances from the end.

50. Distance of Rivet From Edge of Plate.—In order to avoid the danger of fracturing the material, rivets should not be placed too near the edge or end of the plate. Owing to the greater strength of wrought iron in the direction of the fiber, however, rivets in wrought iron may be placed nearer the edge than the end. According to practice, the distance between the edge of any piece and the center of the rivet hole should not be less than $1\frac{1}{4}$ inches for $\frac{3}{4}$- and $\frac{7}{8}$-inch rivets, except in bars less than $2\frac{1}{2}$ inches wide; when practicable, this distance should be at least two diameters of the rivet for all sizes and should not exceed eight times the thickness of the plate. A simple rule is to make the end distance twice the diameter of the rivets and the side distance $\frac{1}{4}$ inch less than this.

51. Clearance.—A certain amount of **clearance** is required for tools in forming the heads of rivets in shapes having short flanges. For a $\frac{7}{8}$-inch rivet, the distance from the center of the rivet to the back of the angle should not be less than $1\frac{1}{4}$ inches; and for a $\frac{3}{4}$-inch rivet, this distance should not be less than $1\frac{1}{8}$ inches. The perpendicular distance between the center line of a rivet in one leg of an angle and the top of a rivet head on the other leg should not be less than $1\frac{1}{8}$ inches.

52. Beam Connections.—In Figs. 45, 46, and 47 are shown the details for **I**-beam connections of several sizes used in structural steel work. They are designed so as to

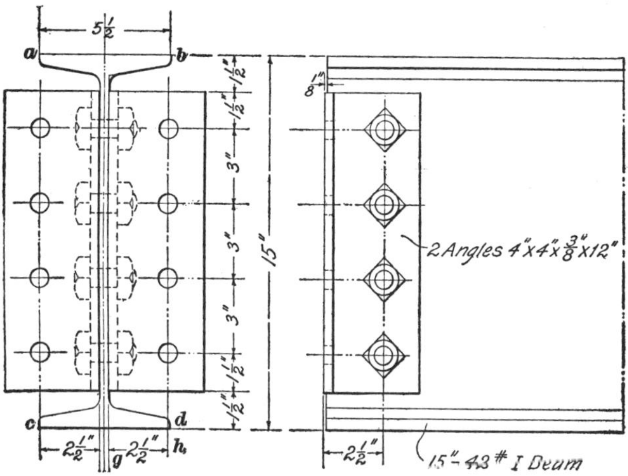

Fig. 45

be strong enough to resist the downward pressure at the ends, due to the safe load supported by the beam, provided the span is not unusually short. In that case, the beam would be able to sustain a greater load than the connection.

Fig. 45 shows *connection angles* bolted to a 15-inch 42-pound beam, the mark # being a conventional sign for pounds. The several dimensions of the beam section *a b c d* are not given, for, in practice, they are determined by the weight of the beam per foot. The necessary dimensions will be found in the catalog of the maker whose beams are to be used, in which the dimensions of the connections for a given size of beam are also given.

The distance $g\,h$, locating the vertical center line of the rivet holes from the back of the angles, is the standard

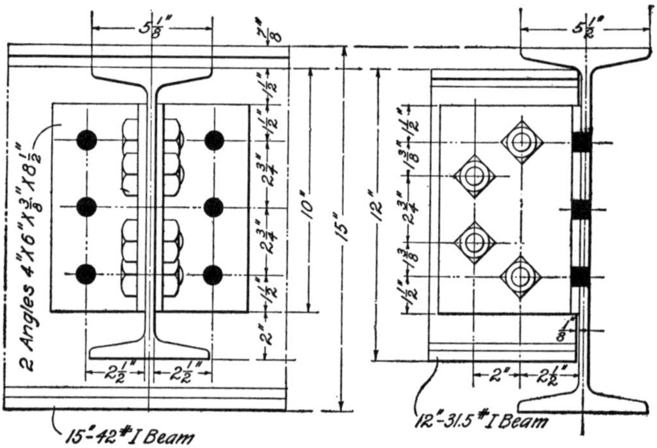

FIG. 46

distance employed by most steel mills for this size of angle. It will be noticed from the right-hand view that the connec-

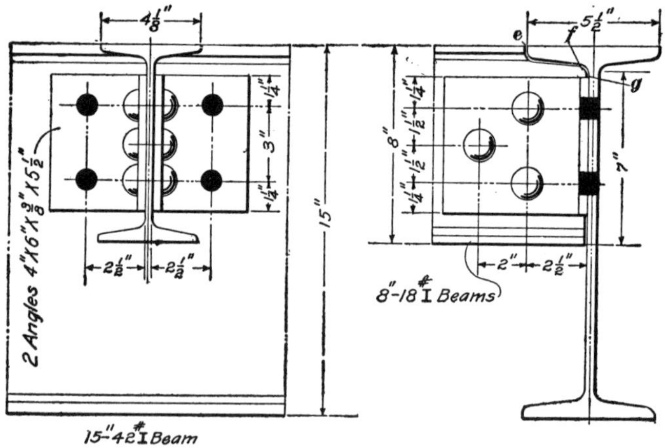

FIG. 47

tion angles are extended outwards $\frac{1}{8}$ inch from the end of the beam, so that when the angles are bolted against the supporting beam, they will fit against it square and not be kept away by the projecting end of the beam, which may be cut somewhat roughly.

Fig. 46 shows the connection of a 12-inch I beam to one 15 inches in depth. This type of joint is always used in the construction of steel-beam floor systems, and is the standard connection for the sizes shown. In Figs. 45 and 46, bolts are shown instead of rivets, but either may be used. The bolt holes shown in solid black in Fig. 46 indicate *field bolts* or *rivets*.

Fig. 47 shows the connection of an 8-inch beam to one 15 inches in depth. This connection differs somewhat from the construction shown in Fig. 46, in that the beams are flush on top. When a beam connection has the flanges of the two beams flush, the auxiliary beam must be cut as shown at *e*, *f*, *g*, so that its web will fit under the flange of the principal beam and the angle connection can be made rigid.

In the notes on the angles indicating their dimensions, as 2 Angles $4'' \times 4'' \times \frac{3}{8}'' \times 12''$, in Fig. 45, the last value 12″ refers to the length of the angle. All the rivets or bolts in these views are $\frac{3}{4}$ inch in diameter.

Note.—When beams are intended for the support of heavy loads and for long spans, they are built up of plates and angles securely riveted together. By this means, the material may be distributed in the most advantageous manner, to obtain the requisite strength with a minimum of weight. Beams constructed in this manner are called *girders*. The long, straight, parallel upper and lower members are called *flanges*, and the vertical plate connecting them is the *web*. If the latter is replaced by braces, the horizontal members are termed *chords*. A chord usually consists of several angles and plates. Chords and braces are shown in Figs. 48 and 49.

53. Rivets in Single Shapes.—There is some difficulty in riveting single shapes, such as bars, angles, etc., to other parts so as to provide proper connections at the ends. Therefore, in order that the efficiency of the metal will be a maximum, the rivets should be arranged so as to make the net cross-sectional area of the bars, etc., as great as possible. The rivets should be placed symmetrically with respect to the axis of the piece, and its cross-section should be reduced as little as possible.

In Fig. 48 (*a*), the cross-section of the plate taken along or parallel with the line *a b* is reduced by only one or two rivet holes. Where the rivets are arranged as shown in Fig. 48 (*b*), a section at right angles to the axis of the mem-

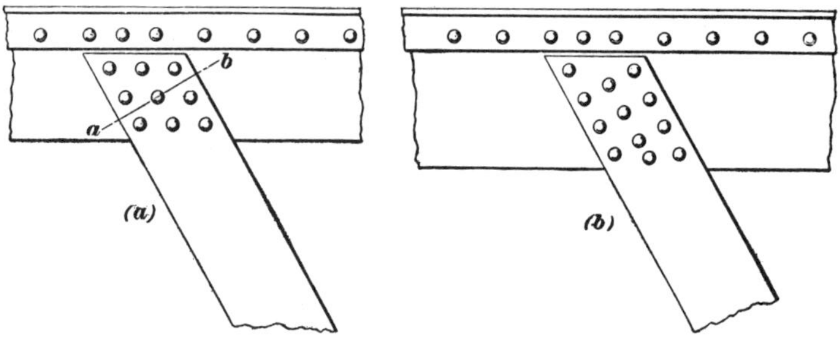

FIG. 48

ber in most cases cuts three rivet holes; consequently, this design is not so efficient as the one shown in (*a*).

To give the greatest strength, angle irons should be attached through both legs, and the rivets should be arranged symmetrically, as in plates. Fig. 49 shows the detail as it is usually made. These joints are all supposed to be in tension.

54. Joints in Compression.—If it were possible to fit the single shapes in Fig. 48 so as to abut against a projecting angle of the horizontal member, a compressive stress could be transmitted entirely by the single bars, and the only use of the rivets would be to hold the parts together. But as a perfect fit cannot be insured, the entire stress is assumed to be transmitted by the rivets, and the calculations are the same as for rivets in tension joints. The only difference is that there is no deduction made from the cross-section for rivet holes, as the rivet completely fills the hole and transmits the stress.

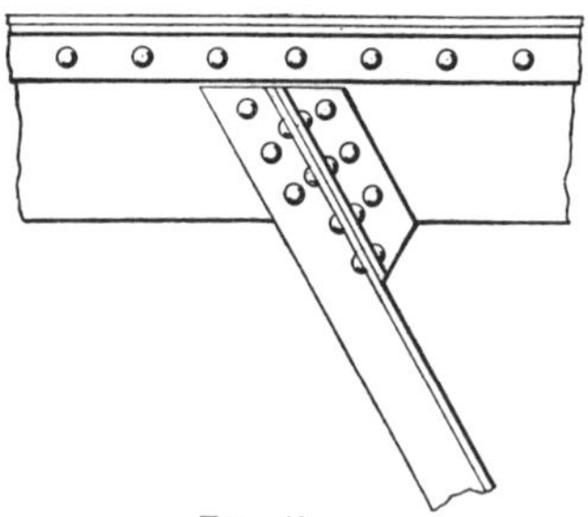

FIG. 49

Where several bars under compression are joined together to transmit the stress, it is necessary sometimes to have the

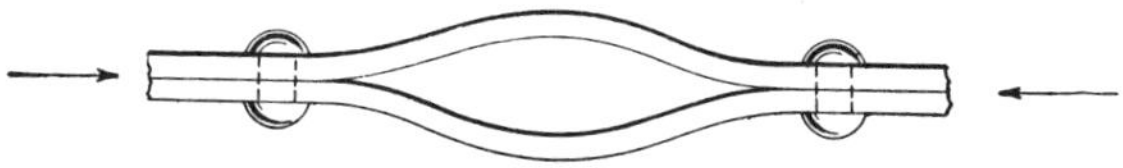

FIG. 50

longitudinal pitch less than in tension joints, so as to avoid the danger of buckling the plates between the rivets, as shown in Fig. 50.

TABLE I

TAPERED KEYS AND KEY SEATS FOR MARINE ENGINES

(Newport News Shipbuilding and Dry Dock Company)

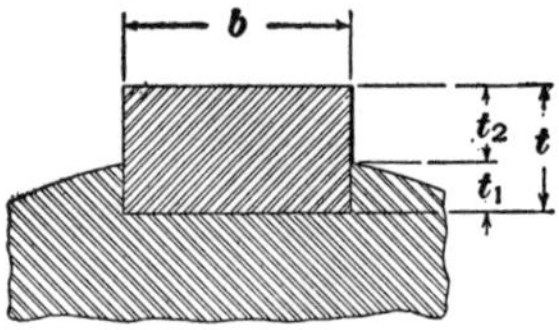

Diameter of Shaft. Inches d	Inch b	Inch t	Inch t_1	Inch t_2	Diameter of Shaft. Inches d	Inches b	Inches t	Inch t_1	Inch t_2
$\frac{1}{2}$	$\frac{7}{32}$	$\frac{3}{16}$	$\frac{1}{16}$	$\frac{1}{8}$	5	$1\frac{1}{16}$	$\frac{5}{8}$	$\frac{1}{4}$	$\frac{3}{8}$
$\frac{5}{8}$	$\frac{1}{4}$	$\frac{3}{16}$	$\frac{1}{16}$	$\frac{1}{8}$	$5\frac{1}{4}$	$1\frac{1}{8}$	$\frac{5}{8}$	$\frac{1}{4}$	$\frac{3}{8}$
$\frac{3}{4}$	$\frac{9}{32}$	$\frac{3}{16}$	$\frac{1}{16}$	$\frac{1}{8}$	$5\frac{1}{2}$	$1\frac{3}{16}$	$\frac{5}{8}$	$\frac{1}{4}$	$\frac{3}{8}$
$\frac{7}{8}$	$\frac{9}{32}$	$\frac{7}{32}$	$\frac{3}{32}$	$\frac{1}{8}$	$5\frac{3}{4}$	$1\frac{3}{16}$	$\frac{11}{16}$	$\frac{5}{16}$	$\frac{3}{8}$
1	$\frac{5}{16}$	$\frac{7}{32}$	$\frac{3}{32}$	$\frac{1}{8}$	6	$1\frac{1}{4}$	$\frac{11}{16}$	$\frac{5}{16}$	$\frac{3}{8}$
$1\frac{1}{8}$	$\frac{11}{32}$	$\frac{1}{4}$	$\frac{3}{32}$	$\frac{5}{32}$	$6\frac{1}{4}$	$1\frac{5}{16}$	$\frac{3}{4}$	$\frac{5}{16}$	$\frac{7}{16}$
$1\frac{1}{4}$	$\frac{3}{8}$	$\frac{1}{4}$	$\frac{3}{32}$	$\frac{5}{32}$	$6\frac{1}{2}$	$1\frac{3}{8}$	$\frac{3}{4}$	$\frac{5}{16}$	$\frac{7}{16}$
$1\frac{3}{8}$	$\frac{3}{8}$	$\frac{1}{4}$	$\frac{3}{32}$	$\frac{5}{32}$	$6\frac{3}{4}$	$1\frac{3}{8}$	$\frac{3}{4}$	$\frac{5}{16}$	$\frac{7}{16}$
$1\frac{1}{2}$	$\frac{13}{32}$	$\frac{1}{4}$	$\frac{3}{32}$	$\frac{5}{32}$	7	$1\frac{7}{16}$	$\frac{13}{16}$	$\frac{5}{16}$	$\frac{1}{2}$
$1\frac{3}{4}$	$\frac{7}{16}$	$\frac{5}{16}$	$\frac{1}{8}$	$\frac{3}{16}$	$7\frac{1}{4}$	$1\frac{1}{2}$	$\frac{13}{16}$	$\frac{5}{16}$	$\frac{1}{2}$
2	$\frac{1}{2}$	$\frac{5}{16}$	$\frac{1}{8}$	$\frac{3}{16}$	$7\frac{1}{2}$	$1\frac{9}{16}$	$\frac{7}{8}$	$\frac{3}{8}$	$\frac{1}{2}$
$2\frac{1}{4}$	$\frac{9}{16}$	$\frac{5}{16}$	$\frac{1}{8}$	$\frac{3}{16}$	$7\frac{3}{4}$	$1\frac{9}{16}$	$\frac{7}{8}$	$\frac{3}{8}$	$\frac{1}{2}$
$2\frac{1}{2}$	$\frac{5}{8}$	$\frac{3}{8}$	$\frac{1}{8}$	$\frac{1}{4}$	8	$1\frac{5}{8}$	$\frac{7}{8}$	$\frac{3}{8}$	$\frac{1}{2}$
$2\frac{3}{4}$	$\frac{5}{8}$	$\frac{3}{8}$	$\frac{1}{8}$	$\frac{1}{4}$	$8\frac{1}{4}$	$1\frac{5}{8}$	$\frac{15}{16}$	$\frac{3}{8}$	$\frac{9}{16}$
3	$\frac{11}{16}$	$\frac{7}{16}$	$\frac{3}{16}$	$\frac{1}{4}$	$8\frac{1}{2}$	$1\frac{3}{4}$	$\frac{15}{16}$	$\frac{3}{8}$	$\frac{9}{16}$
$3\frac{1}{4}$	$\frac{3}{4}$	$\frac{7}{16}$	$\frac{3}{16}$	$\frac{1}{4}$	$8\frac{3}{4}$	$1\frac{3}{4}$	$\frac{15}{16}$	$\frac{3}{8}$	$\frac{9}{16}$
$3\frac{1}{2}$	$\frac{13}{16}$	$\frac{7}{16}$	$\frac{3}{16}$	$\frac{1}{4}$	9	$1\frac{3}{4}$	1	$\frac{7}{16}$	$\frac{9}{16}$
$3\frac{3}{4}$	$\frac{13}{16}$	$\frac{1}{2}$	$\frac{3}{16}$	$\frac{5}{16}$	$9\frac{1}{4}$	$1\frac{7}{8}$	1	$\frac{7}{16}$	$\frac{9}{16}$
4	$\frac{7}{8}$	$\frac{1}{2}$	$\frac{3}{16}$	$\frac{5}{16}$	$9\frac{1}{2}$	$1\frac{7}{8}$	$1\frac{1}{16}$	$\frac{7}{16}$	$\frac{5}{8}$
$4\frac{1}{4}$	$\frac{15}{16}$	$\frac{9}{16}$	$\frac{1}{4}$	$\frac{5}{16}$	$9\frac{3}{4}$	2	$1\frac{1}{16}$	$\frac{7}{16}$	$\frac{5}{8}$
$4\frac{1}{2}$	1	$\frac{9}{16}$	$\frac{1}{4}$	$\frac{5}{16}$	10	2	$1\frac{1}{16}$	$\frac{7}{16}$	$\frac{5}{8}$
$4\frac{3}{4}$	1	$\frac{9}{16}$	$\frac{1}{4}$	$\frac{5}{16}$					

NOTE.—Taper of key = $\frac{1}{8}$ inch per foot.

TABLE II

SQUARE KEYS AND KEY SEATS FOR MACHINE TOOLS

(Wm. Sellers & Company)

Diameter of Shaft Inches	Size of Key Inches		Size of Key Seat Inches	
	b	t	b	$\frac{t}{2}$
$\frac{1}{2}$ and under	$\frac{1}{16}$	$\frac{1}{16}$	$\frac{1}{16}$	$\frac{1}{32}$
$\frac{1}{2}$ to $\frac{15}{16}$	$\frac{1}{8}$	$\frac{1}{8}$	$\frac{1}{8}$	$\frac{1}{16}$
1 to $1\frac{3}{16}$	$\frac{3}{16}$	$\frac{3}{16}$	$\frac{3}{16}$	$\frac{3}{32}$
$1\frac{1}{4}$ to $1\frac{7}{16}$	$\frac{1}{4}$	$\frac{1}{4}$	$\frac{1}{4}$	$\frac{1}{8}$
$1\frac{1}{2}$ to $1\frac{11}{16}$	$\frac{5}{16}$	$\frac{5}{16}$	$\frac{5}{16}$	$\frac{5}{32}$
$1\frac{3}{4}$ to $2\frac{3}{16}$	$\frac{7}{16}$	$\frac{7}{16}$	$\frac{7}{16}$	$\frac{7}{32}$
$2\frac{1}{4}$ to $2\frac{11}{16}$	$\frac{9}{16}$	$\frac{9}{16}$	$\frac{9}{16}$	$\frac{9}{32}$
$2\frac{3}{4}$ to $3\frac{15}{16}$	$\frac{11}{16}$	$\frac{11}{16}$	$\frac{11}{16}$	$\frac{11}{32}$
4 to $5\frac{7}{16}$	$\frac{13}{16}$	$\frac{13}{16}$	$\frac{13}{16}$	$\frac{13}{32}$
$5\frac{1}{2}$ to $6\frac{15}{16}$	$\frac{15}{16}$	$\frac{15}{16}$	$\frac{15}{16}$	$\frac{15}{32}$
7 to $8\frac{15}{16}$	$1\frac{1}{16}$	$1\frac{1}{16}$	$1\frac{1}{16}$	$\frac{17}{32}$

NOTE.—The dimensions of keys for shafting, as adopted by Wm. Sellers & Company, may also be derived from this table, if the values for t are increased by $\frac{1}{16}$ inch. In this table, the key seats in the hub and the shaft are of equal depth; some manufacturers prefer to make the depth of the key seat in the shaft equal to $\frac{9}{16}\,t$. In computing, the length of the key seat is one and one-half times the nominal diameter of the shaft.

TABLE III

TAPER PINS

Number	Diameter at Large End Inch	Approximate Fractional Sizes Inch	Number	Diameter at Large End Inch	Approximate Fractional Sizes Inch
0	.156	$\frac{5}{32}$	6	.341	$\frac{11}{32}$
1	.172	$\frac{11}{64}$	7	.409	$\frac{13}{32}$
2	.193	$\frac{3}{16}$	8	.492	$\frac{1}{2}$
3	.219	$\frac{7}{32}$	9	.591	$\frac{19}{32}$
4	.250	$\frac{1}{4}$	10	.706	$\frac{23}{32}$
5	.289	$\frac{19}{64}$			

TABLE IV

PITCH AND EFFICIENCY OF RIVETED JOINTS

Thickness of Plate Inch t	Diameter of Rivet Inch d'	Diameter of Hole Inch d	Pitch Inches h		Efficiency of Joint y	
			Single	Double	Single	Double
$\frac{1}{4}$	$\frac{5}{8}$	$\frac{11}{16}$	2	3	.66	.77
$\frac{5}{16}$	$\frac{11}{16}$	$\frac{3}{4}$	$2\frac{1}{16}$	$3\frac{1}{8}$	.64	.76
$\frac{3}{8}$	$\frac{3}{4}$	$\frac{13}{16}$	$2\frac{1}{8}$	$3\frac{1}{4}$	.62	.75
$\frac{7}{16}$	$\frac{13}{16}$	$\frac{7}{8}$	$2\frac{3}{16}$	$3\frac{3}{8}$	.60	.74
$\frac{1}{2}$	$\frac{7}{8}$	$\frac{15}{16}$	$2\frac{1}{4}$	$3\frac{1}{2}$	.58	.73

TABLE V

STANDARD SINGLE RIVETING FOR TANKS, STACKS, ETC.

(*Newport News Shipbuilding and Dry Dock Company*)

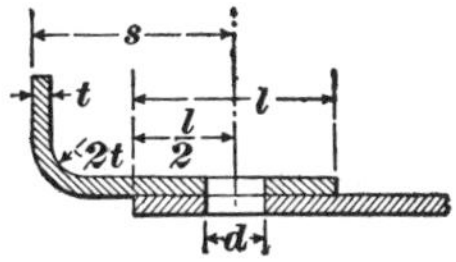

Thickness of Plate Inch t	Diameter of Rivet Inches d	Pitch Inches			Length of Lap Inches l	Distance of Rivet From Side of Tank Inches s
		Oil-Tight Joint	Water-Tight Joint	Non-Water-Tight Joint		
$\frac{1}{16}$	$\frac{1}{4}$	$\frac{13}{16}$	1	2	$\frac{3}{4}$	$\frac{11}{16}$
$\frac{1}{8}$	$\frac{3}{8}$	$1\frac{1}{8}$	$1\frac{1}{4}$	$2\frac{1}{2}$	$\frac{15}{16}$	1
$\frac{3}{16}$	$\frac{3}{8}$	$1\frac{1}{4}$	$1\frac{1}{2}$	3	$1\frac{1}{8}$	$1\frac{1}{4}$
$\frac{1}{4}$	$\frac{1}{2}$	$1\frac{1}{2}$	$1\frac{3}{4}$	4	$1\frac{1}{2}$	$1\frac{5}{8}$
$\frac{5}{16}$	$\frac{5}{8}$	$1\frac{3}{4}$	2	4	$1\frac{1}{2}$	$1\frac{13}{16}$
$\frac{3}{8}$	$\frac{5}{8}$	2	$2\frac{1}{4}$	5	$1\frac{7}{8}$	$2\frac{3}{16}$
$\frac{7}{16}$	$\frac{3}{4}$	$2\frac{1}{8}$	$2\frac{1}{2}$	6	$2\frac{1}{4}$	$2\frac{9}{16}$
$\frac{1}{2}$	$\frac{3}{4}$	$2\frac{1}{4}$	$2\frac{3}{4}$	6	$2\frac{1}{4}$	$2\frac{3}{4}$
$\frac{9}{16}$	$\frac{7}{8}$	$2\frac{1}{2}$	3	7	$2\frac{5}{8}$	3
$\frac{5}{8}$	$\frac{7}{8}$	$2\frac{5}{8}$	$3\frac{1}{8}$	7	$2\frac{5}{8}$	$3\frac{5}{16}$
$\frac{11}{16}$	$\frac{7}{8}$	$2\frac{3}{4}$	$3\frac{1}{4}$	7	$2\frac{5}{8}$	$3\frac{1}{2}$
$\frac{3}{4}$	1	3	$3\frac{1}{2}$	8	3	$3\frac{7}{8}$
$\frac{13}{16}$	1	$3\frac{1}{8}$	$3\frac{3}{4}$	8	3	$4\frac{1}{16}$
$\frac{7}{8}$	1	$3\frac{1}{4}$	$3\frac{7}{8}$	8	3	$4\frac{1}{4}$
$\frac{15}{16}$	$1\frac{1}{8}$	$3\frac{3}{8}$	$4\frac{1}{8}$	9	$3\frac{3}{8}$	$4\frac{5}{8}$
1	$1\frac{1}{8}$	$3\frac{5}{8}$	$4\frac{1}{4}$	9	$3\frac{3}{8}$	$4\frac{13}{16}$

TABLE VI

MAXIMUM SIZE OF RIVETS IN BEAMS, CHANNELS, AND ANGLES

I Beams						Channels			Angles			
Depth of Beam Inches	Weight per Foot Pounds	Size of Rivet Inch	Depth of Beam Inches	Weight per Foot Pounds	Size of Rivet Inches	Depth of Channel Inches	Weight per Foot Pounds	Size of Rivet Inch	Length of Leg Inches	Size of Rivet Inch	Length of Leg Inches	Size of Rivet Inches
3	5.50	3/8	15	42.00	3/4	3	4.00	3/8	3/4	1/4	2 1/2	3/4
4	7.50	1/2	15	60.00	3/4	4	5.25	1/2	1	3/8	2 3/4	3/4
5	9.75	1/2	15	80.00	7/8	5	6.50	1/2	1 1/4	1/2	3	7/8
6	12.25	5/8	18	55.00	7/8	6	8.00	5/8	1 5/16	1/2	3 1/2	1
7	15.00	5/8	20	65.00	1	7	9.75	5/8	1 3/8	1/2	4	1
8	17.75	3/4	20	80.00	1	8	11.25	3/4	1 1/2	1/2	4 1/2	1
9	21.00	3/4	24	80.00	1	9	13.25	3/4	1 3/4	5/8	5	1
10	25.00	3/4				10	15.00	3/4	2	5/8	6	1
12	31.50	3/4				12	20.50	3/4	2 1/4	3/4	7	1
12	40.00	3/4				15	33.00	3/4	2 5/16	3/4		

TABLE VII

LENGTH OF RIVET SHANK REQUIRED FOR UPSETTING AND TO FORM ONE RIVET HEAD

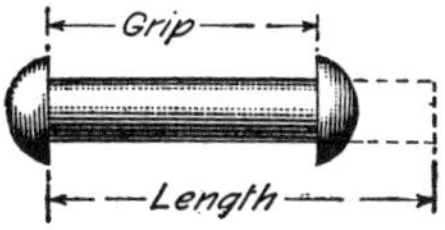

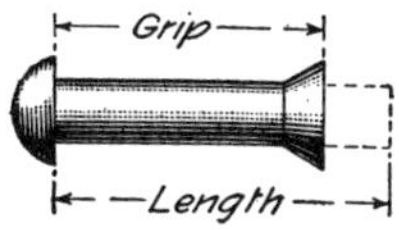

Grip of Rivet Inches	Diameter of Rivet Inches								Grip of Rivet Inches	Diameter of Rivet Inches							
	1/4	3/8	1/2	5/8	3/4	7/8	1	1 1/8		1/4	3/8	1/2	5/8	3/4	7/8	1	1 1/8
1/2	1	1 1/4	1 1/2	1 3/4	1 7/8	2	2 1/8	2 1/4	3 3/8	3 7/8	4 1/4	4 1/2	4 7/8	5	5 1/8	5 1/4	5 3/8
5/8	1 1/8	1 3/8	1 5/8	1 7/8	2	2 1/8	2 1/4	2 3/8	3 1/2	4	4 3/8	4 5/8	5	5 1/8	5 1/4	5 3/8	5 1/2
3/4	1 1/4	1 1/2	1 3/4	2	2 1/8	2 1/4	2 3/8	2 1/2	3 5/8	4 1/8	4 1/2	4 3/4	5 1/8	5 1/4	5 3/8	5 1/2	5 5/8
7/8	1 3/8	1 5/8	1 7/8	2 1/8	2 1/4	2 3/8	2 1/2	2 5/8	3 3/4	4 1/4	4 5/8	4 7/8	5 1/4	5 3/8	5 1/2	5 5/8	5 3/4
1	1 1/2	1 3/4	2	2 1/4	2 3/8	2 1/2	2 5/8	2 3/4	3 7/8	4 3/8	4 3/4	5	5 3/8	5 1/2	5 5/8	5 3/4	5 7/8
1 1/8	1 5/8	1 7/8	2 1/8	2 3/8	2 1/2	2 5/8	2 3/4	2 7/8	4	4 1/2	4 7/8	5 1/8	5 1/2	5 5/8	5 3/4	5 7/8	6
1 1/4	1 3/4	2	2 1/4	2 1/2	2 5/8	2 3/4	2 7/8	3	4 1/8	4 5/8	5	5 1/4	5 5/8	5 3/4	5 7/8	6	6 1/8
1 3/8	1 7/8	2 1/8	2 3/8	2 5/8	2 7/8	3	3	3 1/8	4 1/4	4 3/4	5 1/8	5 1/2	5 3/4	5 7/8	6	6 1/8	6 1/4
1 1/2	2	2 1/4	2 1/2	2 3/4	3	3 1/8	3 1/8	3 1/4	4 3/8	4 7/8	5 1/4	5 5/8	5 7/8	6	6 1/8	6 1/4	6 3/8
1 5/8	2 1/8	2 3/8	2 5/8	2 7/8	3 1/8	3 1/4	3 1/4	3 1/2	4 1/2	5	5 3/8	5 3/4	6	6 1/8	6 1/4	6 3/8	6 1/2
1 3/4	2 1/4	2 1/2	2 3/4	3	3 1/4	3 3/8	3 1/2	3 5/8	4 5/8	5 1/8	5 1/2	5 7/8	6 1/8	6 1/4	6 3/8	6 1/2	6 5/8
1 7/8	2 3/8	2 5/8	2 7/8	3 1/4	3 3/8	3 1/2	3 5/8	3 3/4	4 3/4	5 1/4	5 5/8	6	6 1/4	6 1/2	6 5/8	6 3/4	6 3/4
2	2 1/2	2 3/4	3 1/8	3 3/8	3 1/2	3 5/8	3 3/4	3 7/8	4 7/8	5 3/8	5 3/4	6 1/8	6 1/2	6 5/8	6 3/4	6 7/8	6 7/8
2 1/8	2 5/8	2 7/8	3 1/4	3 1/2	3 5/8	3 3/4	3 7/8	4	5	5 1/2	5 7/8	6 1/4	6 5/8	6 3/4	6 7/8	7	7
2 1/4	2 3/4	3	3 3/8	3 5/8	3 3/4	3 7/8	4	4 1/8	5 1/8	5 5/8	6	6 3/8	6 3/4	6 7/8	7	7 1/8	7 1/8
2 3/8	2 7/8	3 1/8	3 1/2	3 3/4	3 7/8	4	4 1/8	4 1/4	5 1/4	5 3/4	6 1/8	6 1/2	6 7/8	7	7 1/8	7 1/4	7 1/4
2 1/2	3	3 1/4	3 5/8	3 7/8	4	4 1/8	4 1/4	4 3/8	5 3/8	5 7/8	6 1/4	6 5/8	7	7 1/8	7 1/4	7 3/8	7 3/8
2 5/8	3 1/8	3 1/2	3 3/4	4	4 1/8	4 1/4	4 3/8	4 1/2	5 1/2	6	6 3/8	6 3/4	7 1/8	7 1/4	7 3/8	7 1/2	7 1/2
2 3/4	3 1/4	3 5/8	3 7/8	4 1/8	4 1/4	4 3/8	4 1/2	4 5/8	5 5/8	6 1/8	6 1/2	6 7/8	7 1/4	7 3/8	7 1/2	7 5/8	7 5/8
2 7/8	3 3/8	3 3/4	4	4 1/4	4 3/8	4 1/2	4 5/8	4 3/4	5 3/4	6 1/4	6 3/4	7	7 3/8	7 5/8	7 5/8	7 3/4	7 3/4
3	3 1/2	3 7/8	4 1/8	4 3/8	4 1/2	4 5/8	4 3/4	4 7/8	5 7/8	6 3/8	6 7/8	7 1/8	7 1/2	7 3/4	7 3/4	7 7/8	7 7/8
3 1/8	3 5/8	4	4 1/4	4 1/2	4 3/4	4 3/4	5	5	6	6 1/2	7	7 1/4	7 5/8	7 7/8	7 7/8	8	8 1/8
3 1/4	3 3/4	4 1/8	4 3/8	4 3/4	4 7/8	5	5 1/8	5 1/4									

Amount, in Inches, to Be Subtracted From Above Lengths for Countersunk Heads

	1/8	1/4	1/2	1/2	5/8	3/4	7/8	7/8		1/8	1/4	1/2	1/2	5/8	3/4	7/8	7/8

TABLE VIII

RIVET SPACING

Size of Rivet Inch	Minimum Pitch Inches	Maximum Pitch at Ends of Compression Members Inches	Maximum Pitch in Flanges of Chords and Girders Inches	Distance From End of Piece to Center of Rivet Inches	
				Minimum	Usual
$\frac{1}{4}$	$\frac{3}{4}$				
$\frac{3}{8}$	$1\frac{1}{8}$				
$\frac{1}{2}$	$1\frac{1}{2}$				
$\frac{5}{8}$	$1\frac{7}{8}$	$2\frac{1}{2}$	4	$\frac{15}{16}$	$1\frac{1}{4}$
$\frac{3}{4}$	$2\frac{1}{4}$	3	4	$1\frac{1}{8}$	$1\frac{1}{2}$
$\frac{7}{8}$	$2\frac{5}{8}$	$3\frac{1}{2}$	4	$1\frac{5}{16}$	$1\frac{3}{4}$
1	3	4	4	$1\frac{1}{2}$	2

MACHINE DESIGN

(PART 3)

DESIGN OF JOURNALS AND BEARINGS

JOURNALS

END JOURNALS

1. Journals are the cylindrical portions of rotating pieces that turn within bearings and form the supports. Journals situated at or near the ends of a shaft, axle, or other rotating piece are termed *end journals*. Any journal situated between two end journals is called a *neck journal*.

The ordinary form of an **end journal** is shown in Fig. 1, and consists simply of a turned down part of the shaft, shoulders

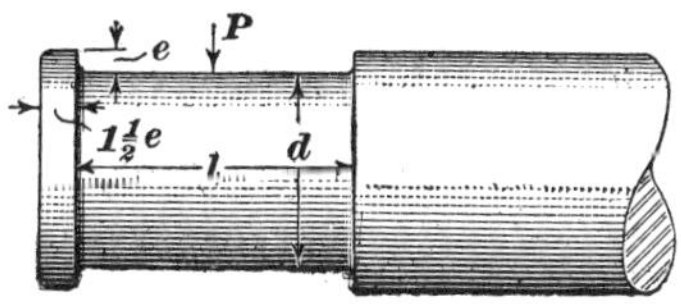

Fig. 1

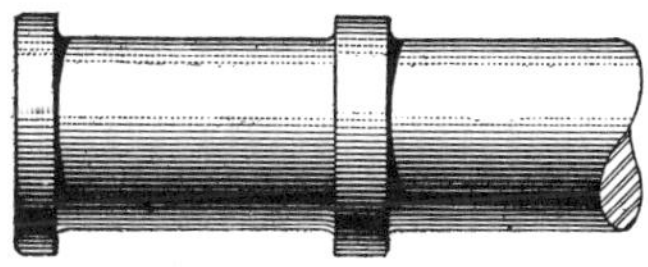

Fig. 2

being produced in this manner at each end of the journal, to prevent end play in the bearing. Sometimes, the same result is obtained without reducing the diameter of the shaft by forging collars at each end of the journal, as shown in Fig. 2.

The length of the box, or brass, on which the journal rests, however, is often made slightly shorter than the journal, permitting a slight motion lengthwise, and thereby securing uniform wear.

SIZE AND PROPORTIONS OF END JOURNALS

2. The chief element in the design of a journal moving slowly or intermittently is strength. When journals run constantly at considerable velocity, strength is not so important a consideration as durability and freedom from liability to heat.

The dimensions of an end journal to give sufficient strength may be calculated by considering the journal as a cantilever uniformly loaded.

Let l = length of journal, in inches;
d = diameter of journal, in inches;
P = total load on journals, in pounds;
S_4 = *safe* stress of material in flexure.

Then, from *Strength of Materials*, the bending moment is $\frac{w l^2}{2} = \frac{P l}{2}$, and the resisting moment is $\frac{S_4 I}{c}$, which for a circular section is

$$S_4 \frac{\frac{\pi d^4}{64}}{\frac{1}{2} d} = S_4 \frac{\pi d^3}{32}$$

Hence,
$$\frac{P l}{2} = S_4 \frac{\pi d^3}{32} \qquad (1)$$

or,
$$d = \sqrt{\frac{16 P}{\pi S_4} \times \frac{l}{d}} = 2.26 \sqrt{\frac{P}{S_4} \times \frac{l}{d}} \qquad (2)$$

Formula **2** gives the diameter of the journal when the ratio $\frac{l}{d}$ is assumed.

Example.—Find the diameter and length of a wrought-iron journal on which there is a load of 1,200 pounds. Assume S_4 = 8,500 pounds and $\frac{l}{d} = 1.4$.

Solution.—Using formula **2**,

$$d = 2.26 \sqrt{\frac{1,200}{8,500} \times 1.4} = 1 \text{ in., nearly. Ans.}$$
$$l = 1.4\, d = 1.4, \text{ say } 1\tfrac{7}{16} \text{ in. Ans.}$$

3. The **bearing surface,** or **projected area,** of a journal is the length multiplied by the diameter; that is, it is the area of the projection of the journal on a plane perpendicular to the direction of the load P. The total load on the journal divided by the projected area gives the pressure per square inch of projected area—a quantity that will be denoted by p.

Hence, $$p = \frac{P}{l\,d}, \text{ or } P = p\,l\,d \qquad (1)$$

In order that the journal may not heat, the pressure p must not exceed a certain limit determined by experience. When this pressure is too great, the oil used to lubricate the journal is squeezed out, and the journal heats rapidly.

From formula **1**, Art. **2**,

$$P = \frac{\pi S_4 d^3}{16\,l}$$

From formula **1**, $$P = p\,l\,d$$

Hence, $$\frac{\pi S_4 d^3}{16\,l} = p\,l\,d, \text{ or } \pi S_4 d^2 = 16\,p\,l^2$$

and $$\frac{l}{d} = \sqrt{\frac{\pi S_4}{16\,p}} \qquad (2)$$

Substituting this value of $\frac{l}{d}$ in formula **2**, Art. **2**, and reducing,

$$d = \sqrt{\frac{4\,P}{\sqrt{\pi S_4 p}}} = 1.5\sqrt{\frac{P}{\sqrt{S_4 p}}} \qquad (3)$$

From formula **1**,

$$l = \frac{P}{d\,p} \qquad (4)$$

Formula **3** may be used to compute the diameter of a journal when the pressure p per square inch of projected area is fixed. The length may then be obtained from formula **4**.

EXAMPLE.—Compute the length and diameter of a steel journal sustaining a load of 12,000 pounds. The safe stress S_4 is 14,000 pounds, and the pressure per square inch of projected area is not to exceed 750 pounds.

SOLUTION.—Using formula **3**,

$$d = 1.5\sqrt{\frac{12{,}000}{\sqrt{14{,}000 \times 750}}} = 2.89, \text{ say } 2\tfrac{7}{8}, \text{ in. Ans.}$$

Hence, $$l = \frac{P}{d\,p} = \frac{12{,}000}{2\frac{7}{8} \times 750} = 5\tfrac{9}{16} \text{ in. Ans.}$$

TABLE I

PRESSURE ON BEARINGS AND SLIDES

Kind of Journal Bearing	Pressure p per Square Inch of Projected Area Pounds
Bearings subject to intermittent load at slow speeds, such as crankpins of shearing machines	3,000
Crosshead neck journals; motion oscillating	800 to 1,400
Crosshead neck journals in locomotives	1,500 to 2,000
Crankpins (slow engines)	800 to 900
Crankpins (fast engines)	500 to 800
Crankpins (marine engines)	400 to 500
Crankpins (locomotives)	1,200 to 1,800
Crankpins (small land engines)	150 to 200
Main crank-shaft bearings (slow)	300 to 450
Main crank-shaft bearings (fast)	200 to 300
Locomotive driving-axle journal	180 to 350
Flywheel-shaft bearings	150 to 250
Eccentric straps	80 to 100
Railway-car journals	250 to 300
Locomotive and tender axle bearings	200 to 250
Line shafting on cast-iron steps (Sellers)	50
Pivots	700
Collar-thrust bearings (slow)	70 to 80
Collar-thrust bearings (fast)	50 to 70
Slides, cast iron on Babbitt metal	200 to 300
Slides, cast iron on cast iron (slow)	60 to 100
Slides, cast iron on cast iron (fast)	40
Steel or iron shaft on lignum vitæ; water lubrication	350

4. The pressure p per square inch of projected area may be taken at from 400 to 800 pounds when the journal runs constantly at a speed under 150 revolutions per minute. For journals that run slowly or intermittently, p may be much greater, while for journals that run faster than 150 revolutions per minute, the pressure p should vary inversely as the number of revolutions. That is, letting N = number of revolutions per minute, $p = \frac{a}{N}$, where a is a constant.

Another consideration affecting the allowable pressure p is the direction of the load. In some journals the load acts only in one direction, generally downwards; in others, as, for example, crankpins and crosshead pins, the direction of the load changes at every revolution. In the latter case, the pressure p may be twice as great as where the load acts in one direction only, because the change in the direction of the load permits a more perfect lubrication of the bearing. When, however, the direction of the load is variable, the safe stress S_4 must be taken smaller than when the direction is constant.

5. The values of p for different kinds of journals, according to Unwin, are given in Table I.

For journals not given in Table I, the value of p must be determined by the judgment of the designer. The value adopted should seldom exceed 750 pounds. For journals running faster than about 150 revolutions per minute, p

TABLE II

SAFE WORKING STRESSES ON END JOURNALS

Direction of Load Constant		Direction of Load Variable	
Material	Safe Working Stress S_4	Material	Safe Working Stress S_4
Steel	14,000	Steel	12,000
Wrought iron .	8,500	Wrought iron .	7,000
Cast iron . . .	4,000	Cast iron . . .	3,000

should vary inversely as the number of revolutions. For example, if a given journal is allowed a bearing pressure of 300 pounds at 150 revolutions, it should be allowed a bearing pressure of only $\frac{300 \times 150}{250} = 180$ pounds if it is required to run at 250 revolutions.

6. The permissible working stress S_4 may be taken as given in Table II.

NECK JOURNALS

7. A **neck journal** similar to an engine crosshead pin may be considered as a beam supported at both ends and uniformly loaded. Consequently, from *Strength of Materials*, the bending moment is $\frac{w\,l^2}{8} = \frac{P\,l}{8}$.

Then,

$$\frac{P\,l}{8} = \frac{S_4 \pi d^3}{32} \text{ (see formula 1, Art. 2)}$$

and

$$d = \sqrt{\frac{4\,P}{\pi\,S_4} \times \frac{l}{d}} = 1.13\sqrt{\frac{P}{S_4} \times \frac{l}{d}} \qquad (1)$$

Formula **1** may be used to calculate a neck journal when $\frac{l}{d}$ is given or assumed.

But, $P = p\,d\,l$ (see formula **1**, Art. **3**)

Hence,

$$\frac{P\,l}{8} = \frac{p\,d\,l^2}{8} = \frac{S_4 \pi d^3}{32}, \text{ or } \frac{l}{d} = \sqrt{\frac{\pi\,S_4}{4\,p}} = .886\sqrt{\frac{S_4}{p}} \qquad (2)$$

Substituting this value of $\frac{l}{d}$ in formula **1**,

$$d = 1.06\sqrt{\frac{P}{\sqrt{p}\,S_4}} \qquad (3)$$

$$l = \frac{P}{p\,d} \qquad (4)$$

From $P = p\,d\,l$,

The same values for S_4 and p may be taken as for end journals.

By comparing formula **3** with formula **3**, Art. **3**, it will be seen that for the same load the neck journal need be only about two-thirds the diameter of the end journal. A comparison of formula **2** with formula **2**, Art. **3**, shows that the ratio $\frac{l}{d}$ in a neck journal is double what it would be in an end journal.

NOTE.—In using formula **3**, Art. **3**, and formula **3**, Art. **7**, it will be found most convenient to use logarithms. Thus, applying logarithms to formula **3**, Art. **3**, it becomes

$$\log d = \log 1.5 + \tfrac{1}{2}[\log P - \tfrac{1}{2}(\log p + \log S_4)]$$

EXAMPLE.—Find the length and diameter of a wrought-iron neck journal, the load being 9,600 pounds and variable in direction. Allow a bearing pressure of 600 pounds per square inch.

SOLUTION.—Using formula **3**,

$$d = 1.06\sqrt{\frac{9{,}600}{\sqrt{600 \times 7{,}000}}}$$

or, $\log d = \log 1.06 + \tfrac{1}{2}[\log P - \tfrac{1}{2}(\log p + \log S_4)]$
$= .02531 + \tfrac{1}{2}[3.98227 - \tfrac{1}{2}(2.77815 + 3.84510)] = .36063$

Whence $d = 2.294 = 2\tfrac{19}{64}$, or $2\tfrac{5}{16}$, in., nearly. Ans.

$$l = \frac{P}{pd} = \frac{9{,}600}{600 \times 2\tfrac{5}{16}} = 7 \text{ in., nearly. Ans.}$$

8. Sometimes, simply as a matter of taste, it may be desirable to make the length of a journal greater than the calculated value. To give the journal the same strength, its diameter must be increased in the proportion given by the following formula:

$$\frac{l_1}{l} = \left(\frac{d_1}{d}\right)^3$$

where l and d are the original and l_1 and d_1 the new lengths and diameters, respectively, of the journal.

For example, suppose that a wrought-iron end journal is subjected to a load of 350 pounds. Assuming $\frac{l}{d} = 1.4$, the dimensions of the journal will be

$$d = 2.26\sqrt{\frac{P}{S_4} \times \frac{l}{d}} = 2.26\sqrt{\frac{350}{8{,}500} \times 1.4} = .5426 \text{ inch;}$$

$$l = 1.4\,d = 1.4 \times .5426 = .7596 \text{ inch}$$

Hence, call the diameter $\frac{9}{16}$ inch and the length $\frac{3}{4}$ inch.

Though a journal of these dimensions would be sufficiently stiff and durable for the load, it is rather small to look well. Suppose the length is made double the calculated value; that is, $l_1 = 2l$. Then, from the formula just given,

$$d_1 = d\sqrt[3]{\frac{l_1}{l}} = .5426\sqrt[3]{2} = .6837, \text{ or } \tfrac{11}{16}, \text{ inch, nearly;}$$

$$l_1 = 2l = 2 \times \tfrac{3}{4} = 1\tfrac{1}{2} \text{ inches}$$

FRICTION OF JOURNALS

9. When relative motion takes place between two surfaces in contact with each other, and subjected to more or less heavy pressure, as in a bearing and a journal, a certain amount of energy is expended in overcoming the frictional resistance to motion. This energy is continually transformed into heat; the temperature of the journal and bearing will therefore increase until a point is reached where the heat is dissipated as rapidly as it is produced. Thus, in designing a journal, the question is not alone that of giving it sufficient strength, but also that of so proportioning it that it may freely radiate the heat developed.

10. The work expended in overcoming the friction of a journal is

$$W = .26 f P N d \text{ foot-pounds per minute,}$$

where f is the coefficient of friction, and N the revolutions per minute.

It is therefore apparent that with the same load and speed, the work expended against the resistance of friction is directly proportional to the diameter of the journal.

In order to illustrate how the dimensions of a journal affect its heat-radiating qualities, a comparison will be made between two journals of equal lengths, but having different diameters, one being 2 and the other 4 inches. The load P is the same for both journals, but the unit pressure p on the larger journal will necessarily be one-half that on the smaller. The frictional resistance of both journals will be the same, as the total pressures P on the journals are equal; but the amount of work lost in friction and converted into heat will not be the same. If both journals make the same number of revolutions per minute, the circumferential speed of one is twice that of the other, and as the amount of work performed is the product of the resistance and the distance through which the resistance is overcome, it follows that the energy lost in revolving the larger journal is twice that expended on the smaller one. The temperature attained by both journals will nevertheless be the same, because the 4-inch journal has a radiating surface twice as great as that of the 2-inch one. As should be clearly seen, not only is no advantage derived from an increase in diameter, but, on the contrary, by reason of the enlarged surface, a greater amount of energy is lost in overcoming the frictional resistance. If, on the other hand, the length of the 2-inch journal were doubled, the peripheral speed would remain the same, and likewise the energy lost in friction, but the radiating surface would be doubled.

As another example, take two journals, one 2 inches in diameter and 6 inches long, and the other 4 inches in diameter and 3 inches long. They each have the same projected area; that is, 12 inches. The latter journal, however, requires double the work to overcome friction that the former requires, and, besides, contains twice as much material. Hence, the $2'' \times 6''$ journal is preferable for a steady load. However, the $4'' \times 3''$ journal is $\frac{4^3 \times 6}{2^3 \times 3} = 16$ times as strong as the other, and would be preferred in situations where the load is variable; also, the journal is liable to shocks, as in the case of crankpins of high-speed engines.

11. It must not be inferred from the preceding that an increase in the length of a journal will always overcome its heating tendency, and that in all cases it would be useless to increase the diameter of the journal. A journal running slowly under heavy pressure might heat less by increasing its diameter and thereby reducing the unit bearing pressure. On the other hand, the advantage derived from increasing the length of a journal may be lost by a tendency of the journal to bend, whereby an excess of pressure is produced at one place, followed by a rise in temperature. By reason of the various points to be taken into consideration, it is difficult to give uniform rules regarding the relation between length and diameter of a journal. In general, the diameter is made as small as the requirements for strength will allow, while the length is made sufficient to keep the bearing pressure within the limits found by experience to be safe.

12. The height of the journal collars, Fig. 1, may be

$$e = \frac{d}{10} + \tfrac{1}{8} \text{ inch}$$

The width of the outer collar is $1\frac{1}{2}\,e$. It is good practice to turn the journal with a fillet in the corner, as the shaft or axle is more liable to crack and fracture if the shoulder is turned with a square corner. The fillet may be outlined by a circular arc drawn with a radius equal to $\frac{1}{2}e$.

PIVOTS AND COLLAR JOURNALS

13. Pivot Journals.—A **pivot journal** is shown in Fig. 3. This type differs from an ordinary journal in that the direction of pressure is parallel with the axis of the shaft instead of perpendicular to it. Thus, the bearing area is the area of the end of the pivot; that is, $.7854\,d^2$.

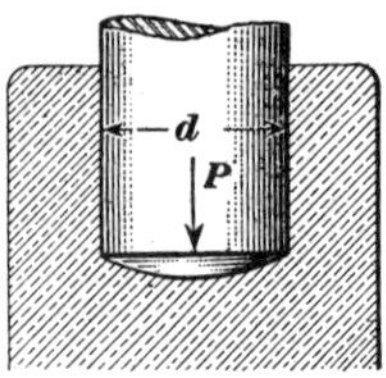

FIG. 3

The diameter of a flat pivot may therefore be found at once by assuming a value for the pressure per square inch of projected area and solving for the area. The formulas in Table III give good results for ordinary cases:

Let P = load on pivot;

d = diameter of pivot;

N = revolutions per minute of pivot.

EXAMPLE.—What should be the diameter of a steel pivot on gun-metal bearings, the load being 200 pounds, and the number of revolutions 320 per minute?

SOLUTION.—

$d = .004\sqrt{PN} = .004\sqrt{200 \times 320} = 1$ in., nearly. **Ans.**

The difficulty met with in pivot journals is the unequal wear of the rubbing surfaces. This is caused by the difference in velocity between parts near the axis of rotation and other parts near the periphery of the journal, the velocity of any particular part being proportional to its distance from the axis of rotation. As the resulting wear of the rubbing surfaces is directly proportional to their respective velocities, it follows that the parts farthest away from the axis of rotation will wear away more quickly. The result is that the

TABLE III

FORMULAS FOR PIVOT JOURNALS

Speed	Wrought-Iron or Steel Pivot on Gun-Metal Bearing	Cast-Iron Pivot on Gun-Metal Bearing	Iron or Steel on Lignum Vitæ Bearing, Moistened With Water	
Case I—Pivot turning very slowly or intermittently . . .	$d = .035\sqrt{P}$	$d = .05\sqrt{P}$	. . . .	**(1)**
Case II—Revolutions per minute less than 150 . . .	$d = .05\sqrt{P}$	$d = .07\sqrt{P}$	$d = .035\sqrt{P}$	**(2)**
Case III—Revolutions per minute more than 150 . .	$d = .004\sqrt{PN}$	. . . . . .	$d = .035\sqrt{P}$	**(3)**

central part of the journal will be gradually exposed to an excess of pressure, followed by abrasion and, possibly, crushing.

14. Collar Journals.—To reduce the extreme differences in velocity between the portions of the bearing surface nearest to and farthest from the centers, it is customary to make use of only the outer part of the rubbing surfaces and to make them ring-shaped. This is the theory on which **collar journals** are based, the aim being to make the collar shallow, thereby avoiding great differences in velocity. In journals exposed to very heavy pressure, the bearing surface is made great enough by increasing the number of collars.

A journal with one collar is shown in Fig. 4, and one with several collars in Fig. 5.

Let d_1 = diameter of collars;
d = diameter of shaft;
b = number of collars;
p = pressure per square inch of projected area;
P = total load, or thrust.

Then, $$.7854\,(d_1^2 - d^2)\,b\,p = P$$

The value of p is usually taken at 60 pounds per square inch for the thrust bearings of propeller shafts, and this value should not be much exceeded in any case.

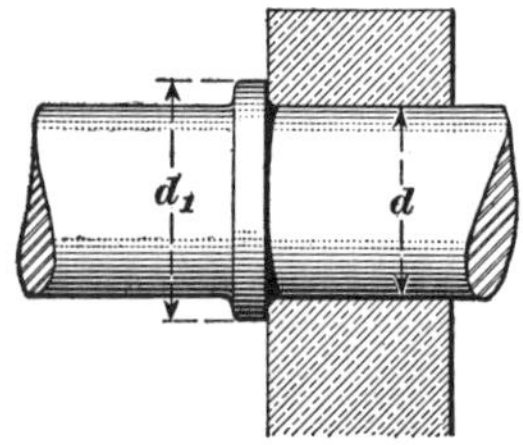

FIG. 4

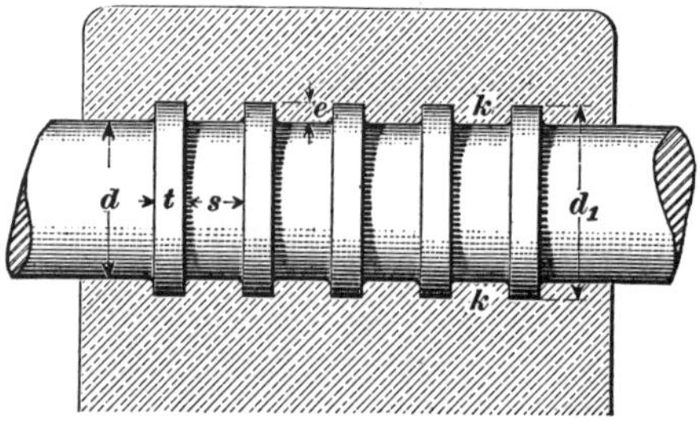

FIG. 5

Then, $$d_1^2 - d^2 = \frac{P}{47\,b}$$

or, $$d_1 = \sqrt{d^2 + \frac{P}{47\,b}}$$

The number of collars b depends on the judgment of the designer. The larger the number, the smaller is the diameter, and the less will be the wear and work of friction. However, when many collars are used, there is danger of bringing all the thrust on one or two.

In Fig. 5, the dimension $e = \frac{1}{2}(d_1 - d)$. Usually the thickness t of the collars is $.8\,e$, and the width s of the space is equal to the thickness t, unless the annular encircling rings k are hollow for water circulation or unless they are lined with white metal. In the latter case, s = 2 to $2\frac{1}{2}\,t$.

EXAMPLES FOR PRACTICE

1. Find the proportions of a cast-iron end journal that turns slowly under a steady load of 15,000 pounds, assuming the length equal to the diameter. Ans. $4\frac{3}{8}$ in. × $4\frac{3}{8}$ in.

2. Find the proportions of a wrought-iron end journal that turns under a load of 8,600 pounds, assuming the safe pressure per square inch of projected area to be 650 pounds. The direction of the load is variable. Ans. 3 in. × $4\frac{7}{16}$ in.

3. Find the dimensions of a steel neck journal, working under a load of 12,000 pounds, direction variable, the allowable bearing pressure to be 1,200 pounds per square inch. Ans. $1\frac{7}{8}$ in. × $5\frac{5}{16}$ in.

4. Find the minimum dimensions of a wrought-iron end journal, the required conditions being that the bearing pressure shall not exceed 750 pounds and that the length shall be one and one-fourth times the diameter. The load on the journal is 9,600 pounds, and variable in direction. Ans. $3\frac{1}{4}$ in. × 4 in.

NOTE.—By using formula **3**, Art. **3**, for solving example 4, the diameter is found to be $3\frac{1}{16}$ inches, nearly. By substituting the value $\frac{l}{1.25}$ for d in formula **4**, Art. **3**, l is found to be equal to 4 inches; but as the quotient $\frac{P}{d\,l}$ in this instance is greater than 750, which is the value of p, the values of d or l must be adjusted to give the desired pressure p. Making $l = 4$ inches, the required value of d may be found by transforming formula **4**, Art. **3**, into the form $d = \frac{P}{l\,p}$, when $d = 3.2$ inches, or $3\frac{1}{4}$ inches, nearly. Ans.

5. Find the diameter of a wrought-iron pivot running in a gun-metal bearing at a speed of 80 revolutions per minute and bearing a load of 800 pounds. Ans. $1\frac{7}{10}$ in.

6. Assuming that five thrust collars on a shaft 10 inches in diameter are subjected to an end thrust of 20,000 pounds, find the diameter and thickness of collars. Ans. $\begin{cases} d_1 = 13.6 \text{ in.} \\ t = 1\frac{7}{16} \text{ in.} \end{cases}$

---

15. Conical Pivot Journals.—When the bearing surface of a pivot journal is conical in shape, as in Fig. 6, it is termed a **conical pivot journal.** Examples of these bearings are found on lathes and similar machines where the end thrust of the journal is not excessive.

The angle u, Fig. 6, of the pivot point does not affect the unit pressure p; this pressure is the same as that of a journal having a flat end whose area is equal to the projection of the conical surface on a plane perpendicular to the axis of the journal. This can be proved in the following manner: In Fig. 6, the load P may be resolved into the two components S, S acting at right angles to the bearing surface, it being supposed that the load is acting in these two directions only. Then,

$$S = \frac{\frac{P}{2}}{\sin\frac{u}{2}}$$

and

$$2S = \frac{P}{\sin\frac{u}{2}} \qquad (1)$$

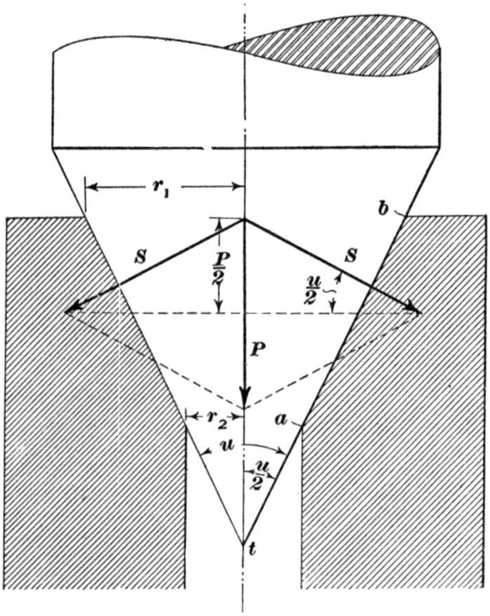

FIG. 6

If A is equal to the surface area of the conical point, then

$$p = \frac{2S}{A}$$

As the active part of the point is the frustum of a cone, the area

$$A = \tfrac{1}{2}(2\pi r_1 + 2\pi r_2)(bt - at)$$

Expressing the lengths bt and at in terms of r_1, r_2, and $\frac{u}{2}$, the formula will be

$$A = \tfrac{1}{2}(2\pi r_1 + 2\pi r_2)\left(\frac{r_1}{\sin\frac{u}{2}} - \frac{r_2}{\sin\frac{u}{2}}\right)$$

or,

$$A = \frac{\pi}{\sin\frac{u}{2}}(r_1^2 - r_2^2) \qquad (2)$$

Inserting this value of A in the formula $p = \frac{2S}{A}$,

$$p = \frac{2S}{\frac{\pi}{\sin\frac{u}{2}}(r_1^2 - r_2^2)}$$

According to formula **1**,

$$2S = \frac{P}{\sin \frac{u}{2}}$$

Therefore,
$$p = \frac{P}{\pi (r_1^2 - r_2^2)}$$

It is seen that the angle u is eliminated from this formula and that it therefore does not affect the unit pressure. The value $\pi (r_1^2 - r_2^2)$ represents the annular surface produced by projecting the conical contact area on a plan perpendicular to the pivot axis.

BEARINGS WITH SLIDING FRICTION

16. Solid Journal Bearings.—The simplest form of bearing for a journal is merely a hole in the frame that supports the rotating piece. Such a bearing is shown in Fig. 7.

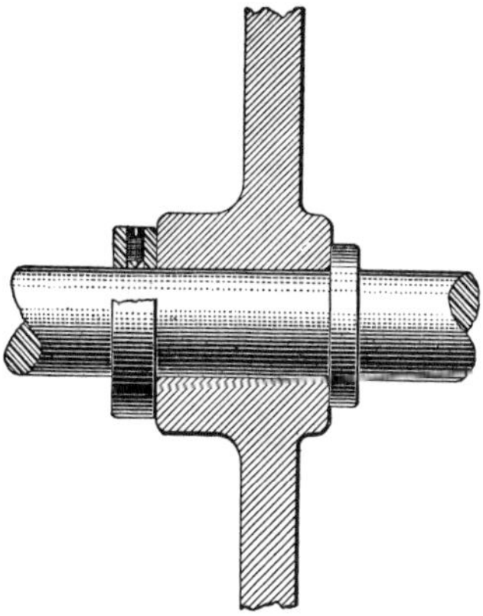

FIG. 7

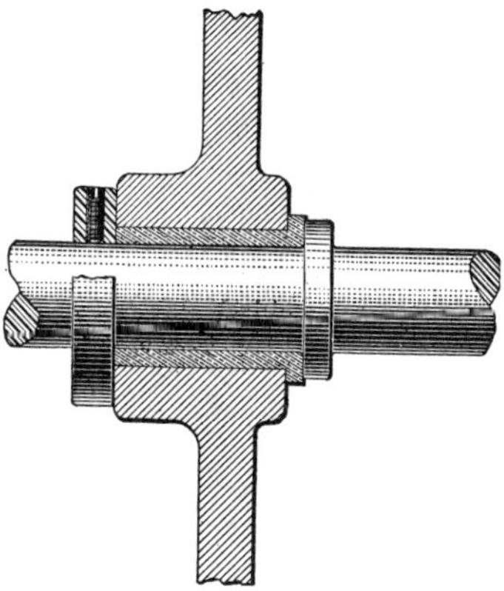

FIG. 8

Motion endwise is prevented by two collars, one of which may be forged on the shaft, and the other made separately and held in place by a setscrew or pin. A boss is cast on the frame, as shown in the figure, in order to give the journal the necessary length and bearing area.

Such a bearing has no means of adjustment to take up the wear; for this reason it is better to use the form of solid bearing shown in Fig. 8. In this bearing the hole is bored out larger than the journal and is lined with a bushing made of brass or other metal. The wear thus comes on the bushing, which can easily be replaced.

17. Divided Bearings.—In many cases, solid bearings are undesirable, and in others it will be impossible to use them. The bearing is then divided, and the parts held together by bolts; when this is done, the parts of the bush-

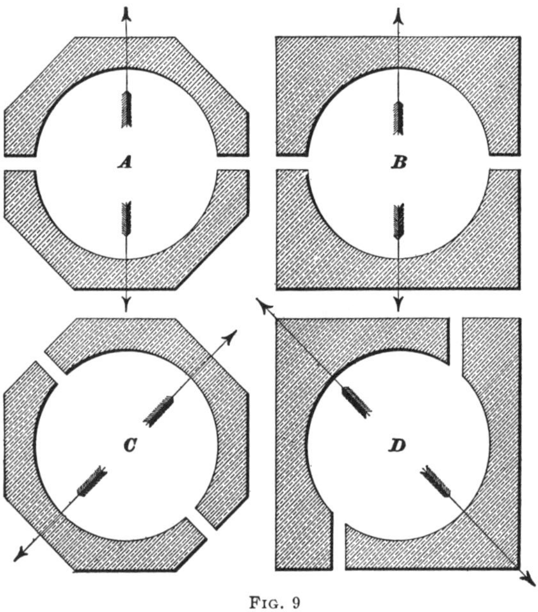

FIG. 9

ing are called *boxes*, or *brasses*. The division of the bearing permits its adjustment for wear.

The load on a bearing usually acts constantly in one direction, and the bearing should be divided so that the line of division is perpendicular, or nearly so, to the direction of the load. Fig. 9 shows various methods of dividing the boxes. At *A* and *B* the direction of the load is vertical; consequently, the boxes are divided horizontally.

Naturally, the wear will come on the top and bottom of the boxes, and the hole will become oval, with the long diameter vertical. The boxes are then screwed closer together until the hole regains its original circular form. The necessity of making the division perpendicular to the direction of the load is thus apparent. At C the direction of the load is oblique, and the adjustment perpendicular to it; at D the direction of the load is oblique also, but the boxes are divided to allow a horizontal adjustment.

In Fig. 10 is shown a simple form of divided bearing; it is made of cast iron, with no separate boxes, but with means of adjustment. The bearing forms part of the frame of the machine. This type of bearing is used only in cheap work. Sometimes, however, the bearing has a recess, or groove,

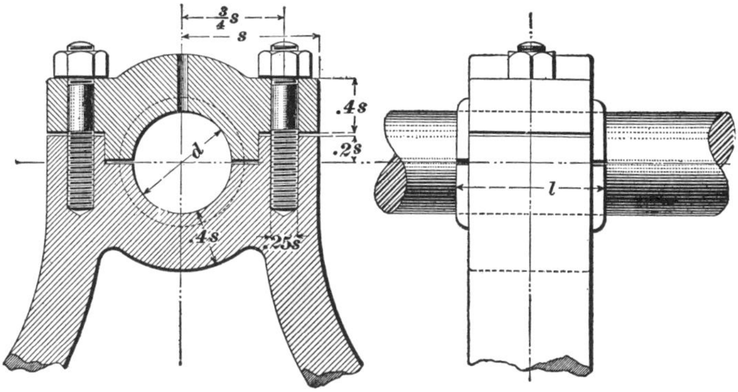

Fig. 10

cast in it, which is filled with Babbitt metal, on which the journal rests. The proportional unit for this bearing is

$$s = 1.15\,d + .4$$

The dimensions l and d of the bearing are the same as those for the journal; the other proportions are given in terms of the unit s, and are shown in the figure.

18. Boxes, or Brasses.—Carefully made bearings are always lined with so-called **boxes**, or **brasses**, which are made of brass, gun metal, phosphor-bronze, white metal, or Babbitt metal, although other alloys are also used. The boxes when worn out may be easily replaced, and being made of softer material than the journal, the latter wears but very little.

An ordinary form of box is shown in Fig. 11. A part of the bearing at each end is made octagonal in cross-section. This part is fitted into an octagonal recess in the *pillow-block*, or *pedestal*, which holds it, and the boxes are thus prevented from turning. Sometimes, the octagonal parts are dispensed with, and the boxes are turned in a lathe, the pillow-blocks, or pedestals, being bored to receive them. It is then necessary to provide the boxes with lugs, or pins, to prevent them from turning.

These boxes may be made according to the following pro-

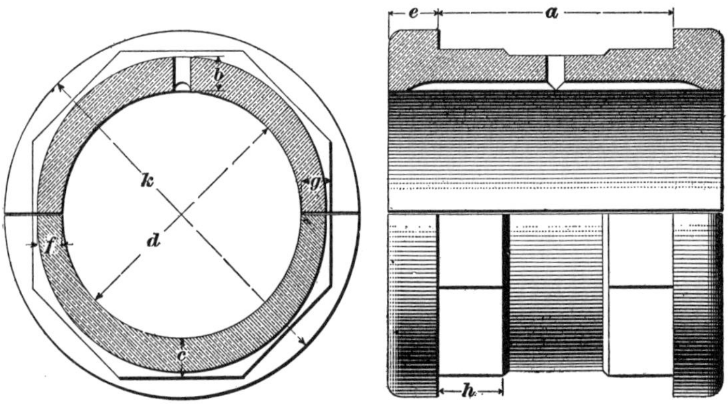

FIG. 11

portions, in which d, the diameter of the journal, is the proportional unit:

$b = \frac{1}{10}d + \frac{3}{16}$ in.;	$g = \frac{1}{8}d$;
$c = \frac{1}{8}d + \frac{3}{16}$ in.;	$h = \frac{1}{4}d$;
$e = \frac{1}{8}d + \frac{1}{8}$ in.;	$k = 1\frac{1}{2}d$;
$f = \frac{1}{10}d$;	d = diameter of journal

The dimension a should be made to fit the pillow-block for which the boxes are intended.

Boxes should be well supplied with grooves and channels, so that oil may be conducted to every part of the journal. It is a good plan to place a groove parallel with the axis of the journal in the upper half of the bearing, or top box, so as to be in direct communication with the oil holes; then, the advancing side of the journal will always carry a thin film of oil along with it. The grooves may be from $\frac{1}{8}$ to $\frac{1}{2}$ inch wide, according to the size of the journal.

19. Lining for Boxes.—Boxes for large bearings are often lined with Babbitt, or some other antifriction, metal. Experience has proved that a bearing will run cooler when so lined, probably because the antifriction metal is softer and thus accommodates itself to the journal more readily than the more rigid metal, mostly brass, of which the boxes are made.

Some of the common methods of lining the boxes are shown in Fig. 12. At (*a*), the Babbitt metal is shown cast

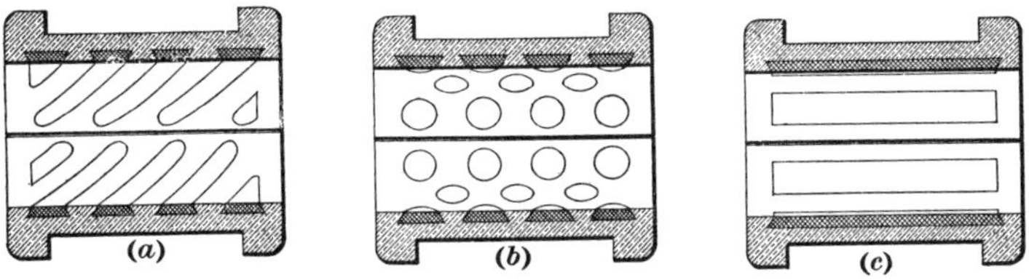

FIG. 12

into shallow helical grooves; at (*b*), into a series of round holes; and at (*c*), into shallow rectangular grooves. Consequently, the journal rests partly on the brass and partly on the Babbitt metal.

In cheap work, the bearing is frequently simply lined with Babbitt metal. A mandrel, the exact size of the journal, is placed inside the bearing, and the melted Babbitt metal is poured around it. In better work, a smaller mandrel is used. After the metal hardens, it is hammered in; the bearing is then bored out to the exact size of the journal.

PEDESTALS

20. The names **pedestal, pillow-block, bearing,** and **journal-box** are applied indiscriminately to one form of bearing, and mean a support for a rotating piece.

21. A form of bearing frequently used for small shafts is shown in Fig. 13. It consists of two parts: the seat that supports the journal, and the cap that is screwed down to the seat. This bearing is lined with Babbitt metal, or, as it is commonly expressed, the bearing is babbitted. The cap is held in place by capscrews—an invariable method in small pedestals.

22. The proportioning of a pedestal is largely a matter of experience. Few or none of the parts are calculated for

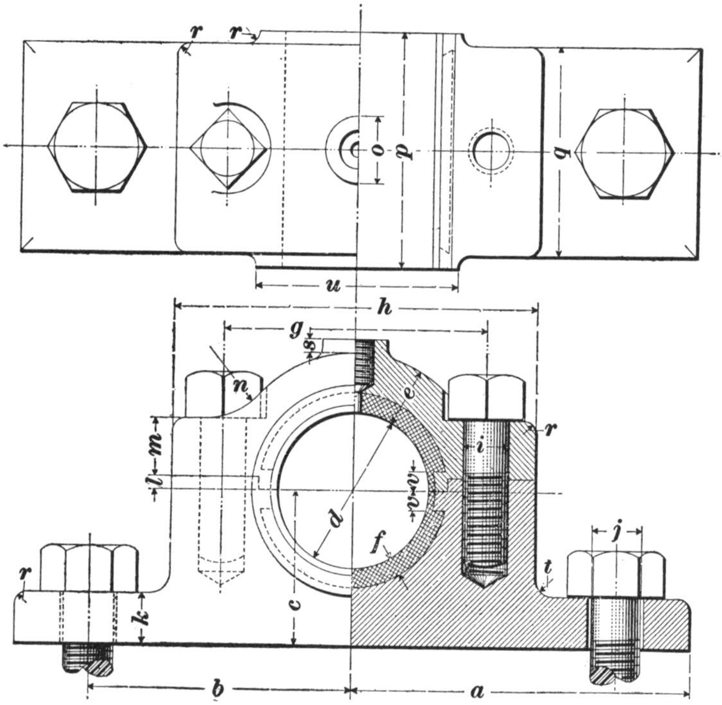

FIG. 13

strength. All the proportions of the pedestals that follow are based on the diameter *d* of the journal as the unit; the length of the bearing is the same as that of the journal.

For the bearing shown in Fig. 13, the following proportions may be used for sizes of journals from $\frac{1}{2}$ to 2 inches in diameter, inclusive. The diameter *d* of the shaft is the unit.

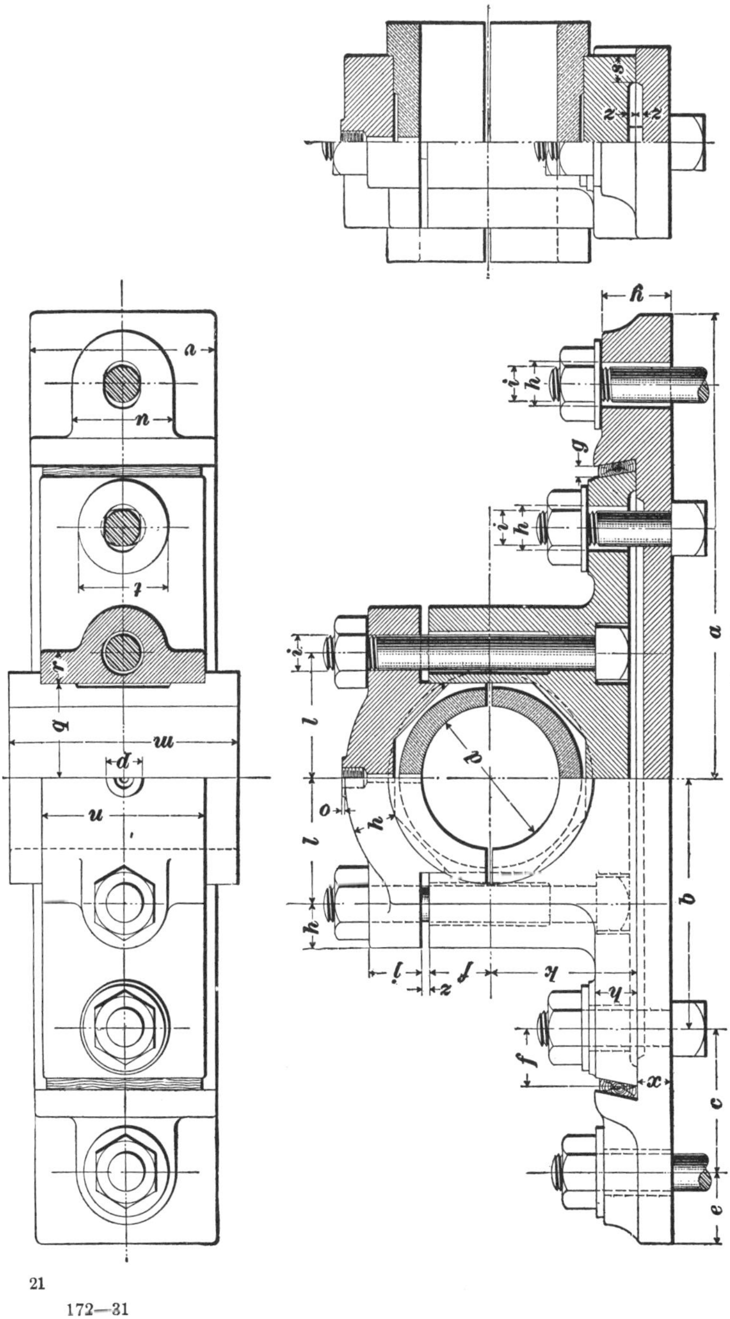

FIG. 14

$a = 2.25\,d$; $l = .08\,d$;
$b = 1.75\,d$; $m = .25\,d + .1875$ in.;
$c = d$; $n = .5\,d$;
$e = .375\,d$; $o = .625$ in. (constant);
$f = .08\,d + .0625$ in.; $p = 1.5\,d$;
$g = 1.75\,d$; $q = 1.333\,d$;
$h = 2.45\,d$; $r = .08\,d$;
$i = .3\,d$; $s = .125$ in. (constant);
$j = .33\,d$; $t = .16\,d$;
$k = .25\,d + .125$ in.; $u = 1.333\,d$;
$v = .125\,d$.

23. In Fig. 14 is shown a common form of pedestal used for somewhat larger journals than the one shown in Fig. 13. This pedestal consists of: (1) a foundation plate that is bolted to the foundation on which the pedestal rests; the plate is essential when the pedestal rests on the brickwork or masonry, but may be dispensed with when the pedestal rests on the frame of the machine; (2) the block that carries the brasses and supports the journal; and (3) the cap that is screwed down over the brasses. The bolt holes in both the foundation plate and the block are oblong, so that the pedestal may be readily adjusted.

The following proportions may be used for this kind of pedestal, with journals from 2 to 6 inches in diameter, inclusive. An oil cup having a $\frac{1}{4}$-inch pipe tap shank may be used on pedestals for journals having diameters from 3 to 4 inches, and $\frac{3}{8}$-inch pipe tap shank for larger sizes up to 6 inches in diameter. The diameter d of the journal is the unit.

NOTE.—The shanks of oil cups and grease cups in the market are made with $\frac{1}{8}$-, $\frac{1}{4}$-, $\frac{3}{8}$-, and $\frac{1}{2}$-inch pipe threads. The amount of oil or grease the cup holds when filled is usually expressed in ounces.

$a = 3.25\,d$; $g = .09\,d$;
$b = 1.75\,d$; $h = .3125\,d$;
$c = d$; $i = .25\,d$;
$e = .5\,d$; $j = .375\,d$;
$f = .4375\,d$; $k = 1.0625\,d$;

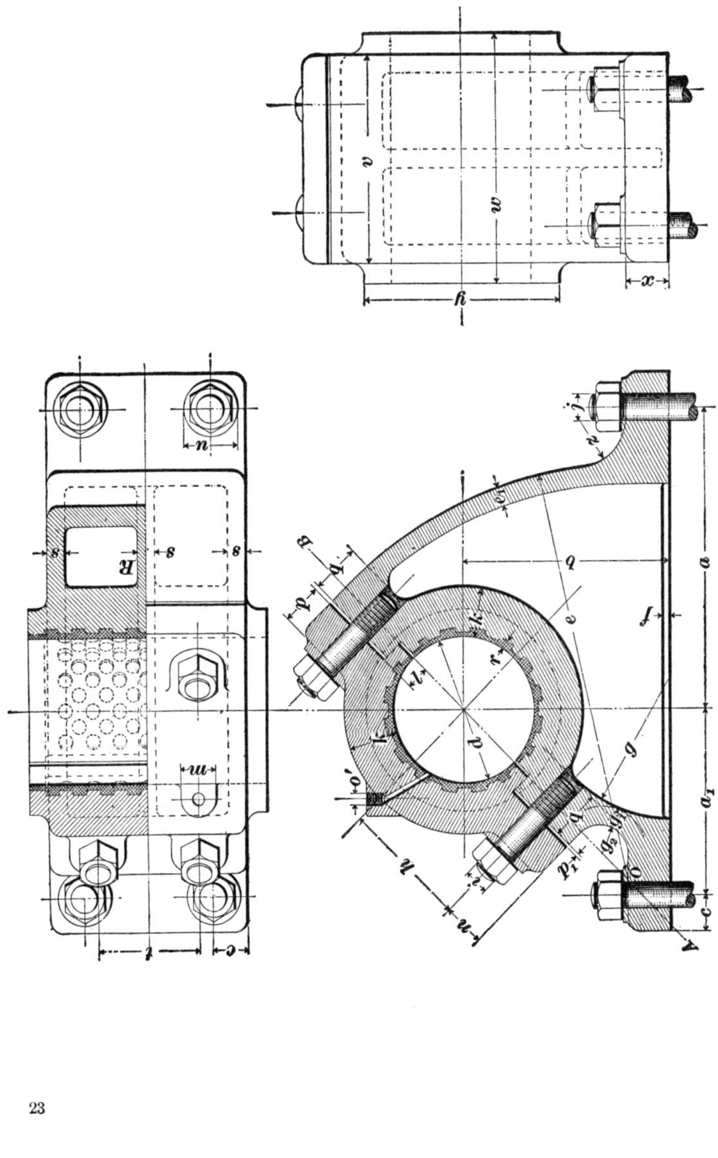

Fig. 15

$l = .875\,d;$	$s = .1875\,d;$
$m = 1.75\,d;$	$t = .65\,d;$
$n = 1.25\,d;$	$u = .75\,d;$
$o = .125$ in. (constant);	$v = 1.375\,d;$
$p = .875$ in. (constant);	$x = .25\,d;$
$q = .625\,d;$	$y = .5\,d;$
$r = .25\,d;$	$z = .0625\,d.$

24. Crank-Shaft Pedestals.—The load on a crank-shaft bearing is due partly to the weight of the shaft and flywheel, and partly to the alternate push and pull of the connecting-rod. The direction of the resultant of these two forces is therefore more or less oblique; consequently, the pedestal is often divided obliquely.

25. A pedestal of the kind just mentioned is shown in Fig. 15, and may be proportioned by taking the diameter d of the journal as the unit and using the following proportions:

$a = 2\,d + 1$ in.;	$n = .3\,d;$
$a_1 = 1.2\,d + 1$ in.;	$p = .25\,d + .25$ in.;
$b = 1.5\,d;$	$p_1 = .04\,d;$
$c = .22\,d + .5$ in.;	$q = .35\,d;$
$e_1 = .1\,d + .25$ in.;	$r = .02\,d + .3125$ in.;
$f = .5$ in. (constant);	$s = .1\,d + .25$ in.;
$g_1 = .1\,d + .25$ in.;	$t = .75\,d;$
$h = .9\,d;$	$u = .27\,d + 1.125$ in.;
$i = .15\,d + .25$ in.;	$v = 1.45\,d;$
$j = .15\,d + .375$ in.;	$w = 1.75\,d;$
$k = .3\,d + .625$ in.;	$x = .25\,d + .625$ in.;
$l = .1\,d + .375$ in.;	$y = 1.3\,d + 1$ in.;
$m = .25\,d;$	$z = .25\,d + 1$ in.

The line AB is at an angle of 45° to the base line. This pedestal, as shown, is babbitted. The Babbitt is held in place and prevented from turning with the journal by providing the surfaces with which the Babbitt comes in contact with round projections, as shown. The projections are about 1 inch in diameter for the larger sizes of pedestals.

To find the radius e, draw a line parallel with, and at a distance x above, the base line. The point of intersection O

of this line with the line AB is the center of the arc having the radius e. The radius g_2 is found by trial. The center of the arc must lie on the line AB. With a radius $g_2 + g_1$, describe an arc from the same center. Draw a line parallel

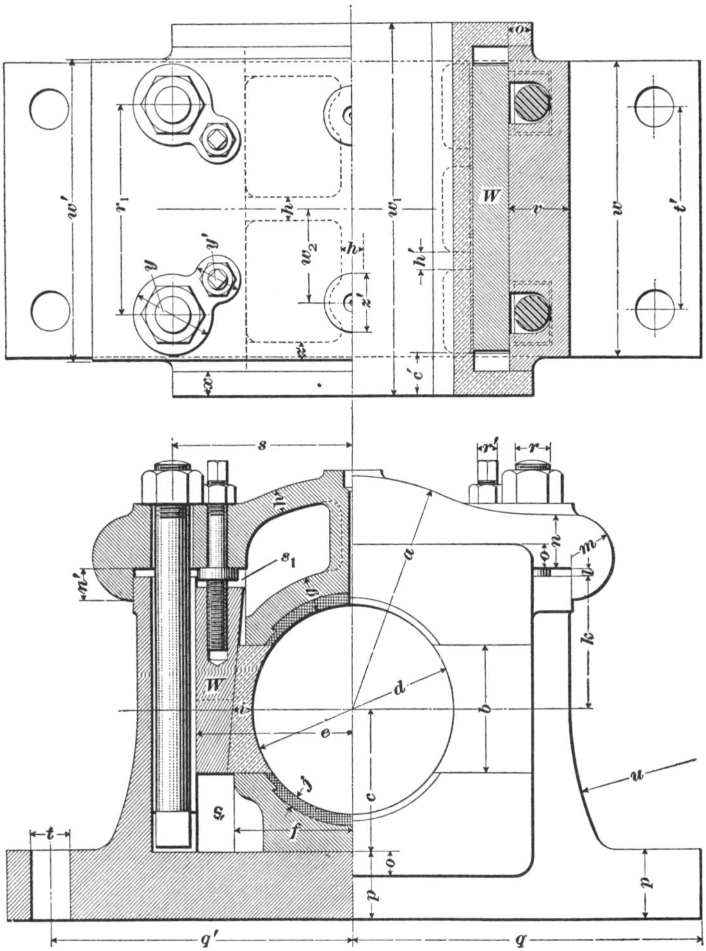

FIG. 16

with the face of the bearing at a distance from it determined by q. The radius g is to be found by trial, the center being on the base line, and the arc tangent to the line determined by q and to the arc determined by $g_2 + g_1$. The rib R may be used on all pedestals in which the diameter of the journal is above 6 inches. For 3- to 4-inch journals, o' may be tapped with a $\frac{1}{4}$-inch pipe tap; up to 6-inch journals, a $\frac{3}{8}$-inch pipe tap; up to 9-inch, a $\frac{1}{2}$-inch pipe tap; and all sizes above 9 inches, a $\frac{3}{4}$-inch pipe tap.

Two oil cups may be used on all bearings in which the diameter of the journal is above 8 inches. This style of pedestal may be used for journals from 3 inches up to 15 inches in diameter, inclusive.

26. Fig. 16 shows a pedestal suitable for the crank-shaft of a horizontal engine with journals from 8 to 20 inches in diameter. The block may be complete in itself, as shown in the figure, but more often it forms part of the engine bed.

The box is in three parts, the cap, the cheek boxes, or cheeks, and the bottom box. The cheeks may be adjusted horizontally by means of wedges W, and the bottom box may be raised by placing packing pieces under it. To obtain the dimensions of this pedestal, use the following proportions, which are based on the unit d, or the diameter of the crank-shaft journal:

$a = d + 1$ in.; $p = .25\,d + .625$ in.;
$b = .5\,d + 1$ in.; $q = 1.75\,d$;
$c = .66\,d$; $q' = 1.5\,d$;
$c' = .2\,d$; $r = .15\,d$;
$e = .825\,d - .25$ in.; $r' = .1\,d$;
$f = .6\,d$; $r_1 = d$;
$g = .1\,d + .5625$ in.; $s = .9\,d$;
$h = .1\,d + .25$ in.; $t = .15\,d + .375$ in.;
$h' = .08\,d$; $t' = .9\,d$;
$i = .11\,d$; $u = 1.5\,d$;
$j = .625$ in. (constant); $v = .25\,d + .375$ in.;
$k = .5\,d + 1.25$ in.; $w = 1.45\,d$;
$l = .375$ in. (constant); $w' = 1.47\,d$;
$m = .175\,d + .3125$ in.; $w_1 = 1.75\,d$;
$n = .25\,d + .25$ in.; $w_2 = \frac{w_1}{4}$;
$n' = .1\,d + .375$ in.;
$o = 1$ in. (constant); $w_3 = w_1 - 2\,x$;

$w_4 = w_1 - 2\,c' + \frac{1}{8}$ in.; $y' = .2\,d + .5$ in.;

$x = .1\,d;$ $z = .09\,d;$

$y = .3\,d + .75$ in.; $z' = 2.5$ in. (constant).

Taper of adjusting wedge $= 1\frac{1}{4}$ inches per foot.

The form and location of the wedge W are shown in the plan view, Fig. 16, and the perspective views, Fig. 17 (b) and (c). It will be noticed that the upper part of the wedge extends across the whole width of the pedestal, until it reaches the flanges of the cap and the cheeks which have corresponding recesses s_1, s_1, Fig. 17 (a) and (b). As shown

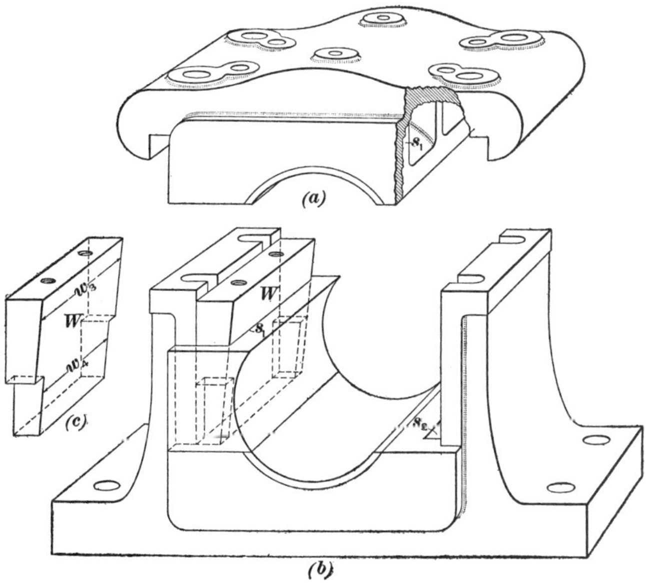

FIG. 17

in Fig. 17 (b), the bottom box has shoulders that make the recess s_2 narrower than those in the cheeks and cap. For this reason, the lower part of the wedge is made narrower, so as to enter the recess s_2. The perspective views will assist in locating the various parts and in understanding the peculiar shape of the cap, the view of the latter being partly broken away in order to show the lower central part.

The dimensions of the wedge may be found in the following manner: Lay off the lengths i and e, Fig. 16, along the horizontal center line of the bearing, the difference between the two dimensions being equal to the thickness of the wedge along this line. Through the left-hand end of dimension i draw a vertical line equal in length to b, the lower end of the line beginning at the lower side of the cheek. As the taper of the wedge is $1\frac{1}{4}$ inches to the foot, calculate the amount of taper for a length equal to b. Draw horizontal lines through the upper and lower ends of this vertical line, and along the upper line set off one-half of the taper to the right of the vertical line, and along the lower one, one-half to the left. By connecting these two points, the taper side of the wedge is determined; this side may now be extended upwards to any point desired, the bottom of the wedge being made flush with the lower side of the cheek.

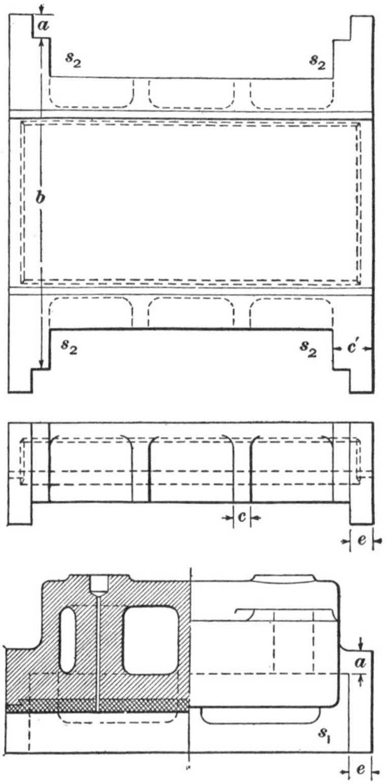

Fig. 18

The center of the circle of diameter y', shown in the plan view, Fig. 16, should be approximately over the center of the adjusting wedge and far enough from the center of the large bolt to permit both nuts to be turned freely.

Further details of the bottom box and the cap are shown in Fig. 18, in which the unit is the same as in Fig. 16, and the proportions are as follows:

a = 1 in. (constant); c' = .2 d;
b = 1.65 d − .5 in.; e = .1 d.
c = .08 d;

The outlines of the dovetailed projections that hold the Babbitt lining in place are indicated by dotted lines.

27. The foundation, or bed casting, is shown in Fig. 19 (*a*), and has dimensions to suit the pedestal shown in Fig. 16. The

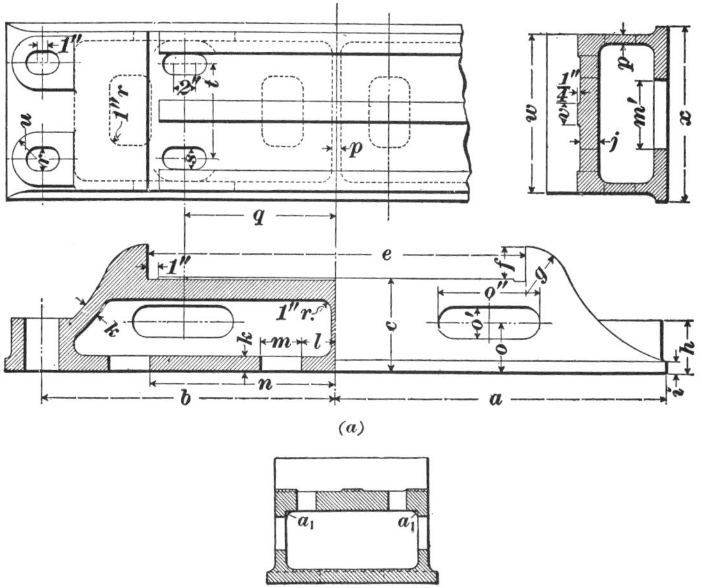

(*a*)

(*b*)

FIG. 19

proportions for this foundation are as follows, the diameter d of the crank-shaft journal being taken as the unit.

a = 2.45 d + 7.25 in.; i = .05 d + .5 in.;
b = 2.3 d + 5.25 in.; j = .05 d + 1.125 in.;
c = .5 d + 3.5 in.; k = .05 d + .75 in.;
e = 3.5 d + 2 in.; l = .25 d + .75 in.;
f = .25 d + .5 in.; m = .4 d;
g = .25 d + 1.75 in.; m' = .6 d;
h = .25 d + 2.25 in.; n = 1.55 d + 2.5 in.;

$o = .25\,d + 2$ in.; $s = .15\,d + .375$ in.;
$o' = .25\,d + .5$ in.; $t = .9\,d$;
$o'' = .5\,d + 4.5$ in.; $u = .15\,d + .875$ in.;
$p = .08\,d$; $v = .2\,d$;
$q = 1.5\,d$; $w = 1.5\,d$;
$r = .15\,d + .375$ in.; $x = 1.65\,d$.

Fig. 19 (b) shows a cross-section taken through two of the holes for the pedestal bolts, and explains the presence

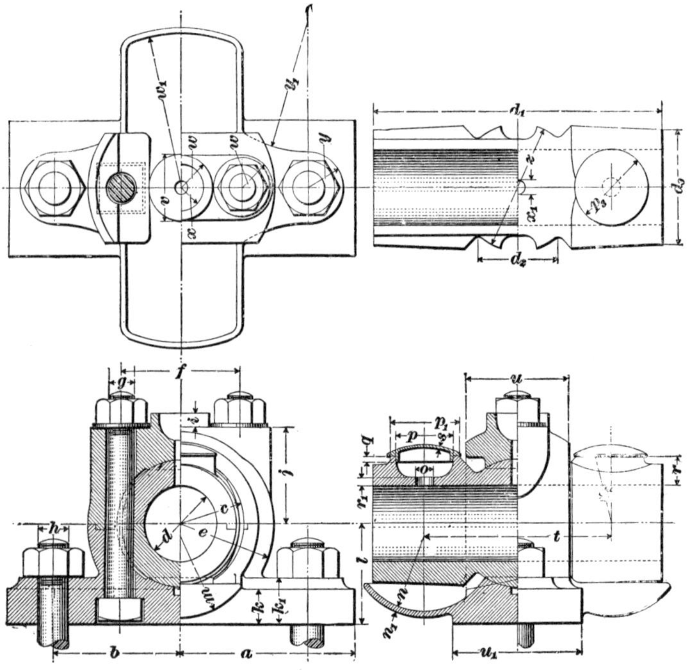

FIG. 20

of the dotted curve at the outside of the holes in the plan view. As shown at a_1, the fillet at the inside of the top has been milled away near the bolt hole, in order to make room for the pedestal bolt head and give the latter a flat bearing surface.

28. Ball-and-Socket Bearings.—The **ball-and-socket bearing** now largely used in the United States was first introduced by William Sellers & Co., of Philadelphia. This bearing has very long boxes made of cast iron bored to fit the journal. In some cases, however, the boxes are cast with a recess in which a Babbitt lining is poured to form a bearing surface for the journal. The boxes have a spherical enlargement at the center, which fits into corresponding hollows in the block and cap, thus making a ball-and-socket joint, which leaves the box free to move slightly in any direction to conform to an inequality or want of alinement in the shaft. When rigid boxes are used, they must necessarily be short on account of unavoidable deflections of the shaft due to belt pull, thrust of gearing, etc. Pivoted boxes, on the contrary, by reason of their flexibility, may be made long, thereby giving a large wearing surface and increased durability.

29. Fig. 20 shows a pedestal with ball-and-socket bearing. The boxes are made of cast iron and have a length equal to four times the diameter of the journal. Ordinarily, the journal is lubricated through the oil hole in the center of the cap, but the top box also has two cups, which are filled with a mixture of tallow and oil that melts at about 100° F. If the bearing becomes heated, this mixture melts, and thus helps in the lubrication.

The dimensions of this bearing are obtained from the following proportions, which are based on the diameter d of the journal as the unit:

$a = 2.375\,d$;	$j = 1.27\,d$;
$b = 1.75\,d$;	$k = .3\,d + .25$ in.;
$c = .875\,d$;	$k_1 = .4\,d + .375$ in.;
$d_1 = 4\,d$;	$l = 1.33\,d$;
$d_2 = 1.1\,d$;	$m = d + .125$ in.;
$d_3 = 1.25\,d + .5$ in.;	$n = d$;
$e = 1.3\,d$;	$n_1 = .125$ in. (constant);
$f = 1.65\,d$;	$o = .25\,d$;
$g = .2\,d + .25$ in.;	$p = .8\,d$;
$h = .2\,d + .375$ in.;	$p_1 = .95\,d$;
$i = .1875\,d$;	$p_2 = d$;

$p_3 = d;$
$q = .07\,d;$
$r = .4\,d;$
$r_1 = .1\,d;$
$s = .0625$ in. (constant);
$t = 2.5626\,d;$
$u = 1.45\,d;$
$u_1 = 1.8\,d;$
$v = .875\,d;$
$w = .2\,d + .375$ in.;
$w_1 = 2.06\,d;$
$x = .18\,d + 2.5$ in.;
$x_1 = .1875\,d;$
$y = .25\,d + .375$ in.;
y_1 to be found by trial;
$z = 1.5\,d + .3125$ in.

WALL BRACKETS AND HANGERS

30. Wall Brackets.—A shaft must sometimes be supported by a bearing fixed to a wall or pillar. In such a case, the bearing is generally supported by a **wall bracket.** It will readily be seen that the bracket is a cantilever with practically a uniform load, due to its own weight, and a concentrated load, due to the weight of the shaft; hence, the bending moment may be determined by the principles of strength of materials.

In order that a cantilever shall be equally strong at all sections, the resisting moment $\frac{S\,I}{c}$ of that section must be proportional to the bending moment at that section. Suppose the bracket in question to be a plate of a constant thickness b and a varying height h. Then, at any section, the bending moment

$$M = \frac{S\,I}{c} = \frac{S\,\frac{1}{12}\,b\,h^3}{\frac{1}{2}\,h} = \frac{1}{6}\,S\,b\,h^2$$

Since $\frac{1}{6}\,S\,b$ remains constant for all sections, h^2 is proportional to the bending moment; or, in other words, the height of a cantilever at any section should be proportional to the square root of the bending moment at that section. It can be shown mathematically that a cantilever of constant width should have a parabolic form when the load is concentrated at the end, and a triangular form when the load is uniformly distributed. In practice, a bracket or other machine part of the cantilever class is given a form that is often neither exactly parabolic nor triangular, but approximates closely

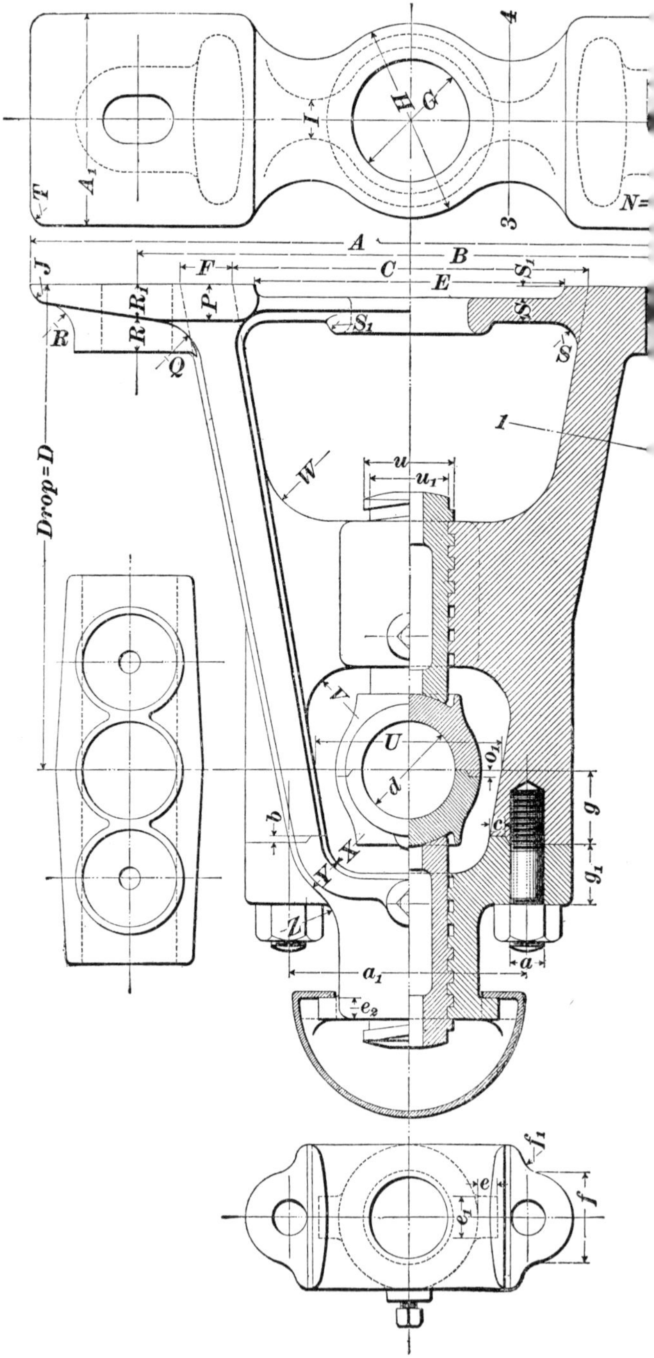

Drop=D
A
B
C
E
F
G
H
I
J
P
Q
R
R_1
S_1
T
U
V
W
X
Y
Z
A_1
a
a_1
b
c
d
e
e_1
e_2
f
f_1
g
g_1
u
u_1
o_1
1
3
4

FIG. 21

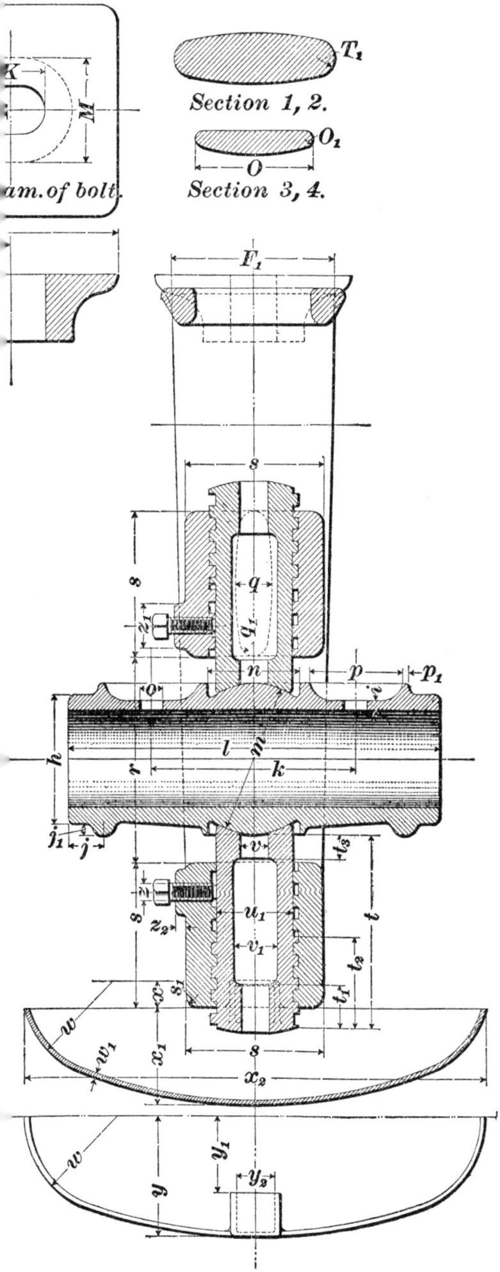

T_1
Section 1, 2.
O_1
O
Section 3, 4.
K
M
am. of bolt.
F_1
s
s
q
q_1
z_1
n
p
p_1
o
i
h
l
m
r
k
j_1
j
v
t_3
z
u_1
s
z_2
v_1
t
t_2
x
s_1
t_1
w
w_1
x_1
s
x_2
w
y_1
y
y_2

to one or the other, according to the taste and judgment of the designer. The bearing of the wall bracket may be made the same as in the hanger shown in Fig. 21.

31. Hangers.—A **hanger** is used when a shaft bearing is to be suspended from the ceiling. Fig. 21 shows a form of a hanger frequently employed.

The frame of the hanger is divided, and the parts are connected by bolts. With such a form the shaft may be more easily removed than when the hanger frame is a solid piece.

The units for determining the leading dimensions of a shaft hanger are the diameter d of the shaft and the drop D of the hanger. The following proportions are suitable for shafts ranging from $1\frac{1}{2}$ inches up to $4\frac{1}{2}$ inches in diameter:

$A = 6\,d + .45\,D$;
$A_1 = 2\,d + .03\,D$;
$B = 4\,d + .35\,D$;
$C = 2\,d + .3\,D$;
$E = 2\,d + .25\,D$;
$F = .5\,d + .01\,D$;
$F_1 = 1.5\,d + .05\,D$;
$G = 1.25\,d$;
$H = 2\,d$;
$I = .4\,d$;
$J = .125\,d + .01\,D$;
$K = .5\,d + .5$ in.;
$L = .25\,d + .5$ in.,
$M = .75\,d + .6875$ in.;
$N = .25\,d + .375$ in.;
$O = 1.25\,d$;
$O_1 = .094\,d + .002\,D$;
$P = .375\,d + .008\,D$;
$Q = .375\,d + .008\,D$;
R and R_1 = (see note);
$S = .25\,d + .005\,D$;
$S_1 = .125\,d + .003\,D$;
$T = .125\,d + .01\,D$;
T_1 = (see note);
$U = 2\,d$;
$V = .5\,d$;
$W = .75\,d$;
$X = .375\,d$;
$Y = .25\,d + .125$ in.;
$Z = .625\,d$;
$a = .15\,d + .375$ in.;
$a_1 = 2.4\,d + .3125$ in.;
$b = .08\,d$;
$c = .125\,d + .0625$ in.;
$e = .2\,d$;
$e_1 = .4\,d$;
$e_2 = .2\,d$;
$f = .375\,d + 1$ in.;
$f_1 = .09\,d + .25$ in.;
$g = .75\,d$;
$g_1 = 1.3125\,d + .125$ in.;
$h = 1.25\,d + .1875$ in.;
$i = .1\,d$;
$j = .25\,d + .25$ in.;
$j_1 = .125\,d + .0625$ in.;
$k = 2.2\,d$;
$l = 4\,d$;
$m = 1.4\,d + .375$ in.;

$n = d$;
$o = .25\,d$;
$o_1 = .0625\,d$;
$p = d$;
$p_1 = .0625\,d$;
$q = .4\,d$;
$q_1 = .15\,d$;
$r = 2.125\,d$;
$s = 1.5\,d$;
$s_1 = .125\,d$;
$t = 2\,d$;
$t_1 = .5\,d$;
$t_2 = d$;
$t_3 = .25\,d$;
$u = .95\,d$;
$u_1 = .85\,d$;
$v = .25\,d + .125$ in.;
$v_1 = .5\,d$;
$w = d$;
$w_1 = .125$ in. (constant);
$x = .25\,d$;
$x_1 = d$;
$x_2 = 4\,d + 2$ in.;
$y = 1.25\,d$;
$y_1 = .75\,d + .0625$ in.;
$y_2 = .4\,d + .0625$ in.;
$z = .06\,d + .25$ in.;
$z_1 = .12\,d + .75$ in.;
$z_2 = .3125$ in. (constant).

Thread of plugs = .5-inch pitch for all sizes.

NOTE.—To find R_1, draw the arc J; also, draw the arc Q tangent to P; then, draw a straight line tangent to these arcs, and R_1 will be the distance along the center line, determined by B included between this tangent and the upper face of the hanger. Having found R_1, make R equal to it.

The radius T_1 is made equal to three-eighths the thickness at the middle.

The boxes of the ball-and-socket bearings shown in Figs. 20 and 21 are made of cast iron, and are bored to fit the journal without lining. The ball, and the recesses in the ends of the plugs into which the ball is fitted, should be finished. The screw threads on the plugs, Fig. 21, may be either cast on the plugs or turned, the latter method being preferable.

The spherical enlargement at the center of the boxes is held between the ends of two hollow cast-iron plugs, which are threaded at their outer ends and screw into bosses cast on the bracket. The boxes may be raised or lowered by screwing the plugs up or down. In order to turn the plugs, the hole at the end is made square to receive a key. When the plugs are in the desired position they are locked by setscrews.

32. Pivot, or Footstep, Bearings.—Bearings that are used to support the ends of vertical shafts are known as **pivot, or footstep, bearings.** An ordinary pivot bearing is shown in Fig. 22. The end of the pivot rotates on a disk *A*, which may be made of steel, brass, or bronze. The brass bushing *B* prevents the pivot from moving sidewise. The end of the pivot should be made of steel, and it may be flat on the end or slightly convex. The proportions are

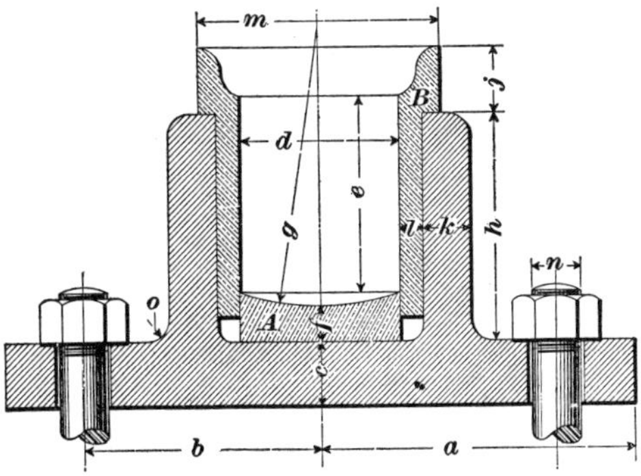

FIG. 22

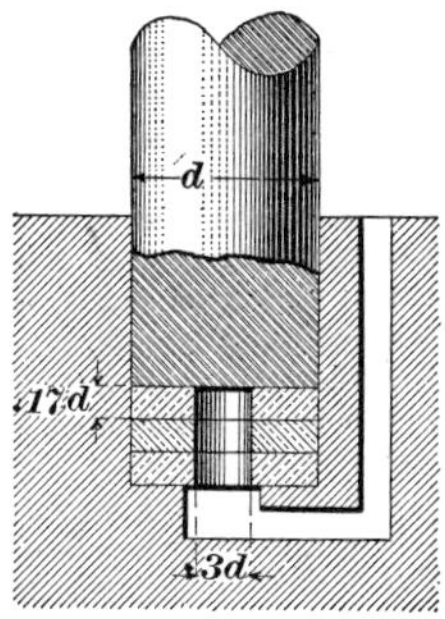

FIG. 23

given in terms of the diameter *d* of the pivot as a unit, and are as follows:

$a = 2\,d$;	$j = .4\,d$;
$b = 1.5\,d$;	$k = .3\,d$;
$c = .25\,d + .375$ in.;	$l = .15\,d$;
$e = 1.25\,d$;	$m = 1.5\,d$;
$f = .2\,d + .125$ in.;	$n = .2\,d + .25$ in.;
$g = 1.75\,d$;	$o = .15\,d$.
$h = 1.4\,d$;	

33. Pivots With Loose Disks.—When pivots are required to run at great speed, the relative speed of the surfaces in contact may be reduced by placing a number of loose disks between the pivot and the footstep, shown in Fig. 23. These disks are sometimes made slightly convex.

If *a* is the number of disks and *n* the revolutions per minute of the pivot, then the relative speed of any of the surfaces in contact is $\frac{n}{a+1}$ revolutions per minute. Suppose, for example, a pivot runs at 3,000 revolutions per minute, and it is desired to reduce the speed between adjacent surfaces to

750 revolutions per minute; then, $750 = \frac{3{,}000}{a+1}$, or $a = 3$, the number of disks required. The pivot and disks may be lubricated by a channel, as shown in the figure. The disks may be made of gun metal or phosphor-bronze. Proportions are given in the figure.

ROLLER AND BALL BEARINGS

ROLLER BEARINGS

34. Rolling Friction.—In Fig. 24, C illustrates a cylinder at the center of which the force P is acting, compelling the cylinder to roll over the surface D and to revolve in the direction shown by the arrow v. As indicated, the surface D undergoes a certain amount of deformation, which may be permanent, if the material is of a plastic nature, or only temporary, if the material is elastic and the compressive stress does not exceed the elastic limit. In the position shown, the point e is lower than b, and in order that the motion may continue, the cylinder must revolve about b as an instantaneous axis of rotation. The moment causing rotation is $P \times a\,b$, and the resisting moment is $W \times b\,c$, W being the load supported by the cylinder. These moments must be equal; hence,

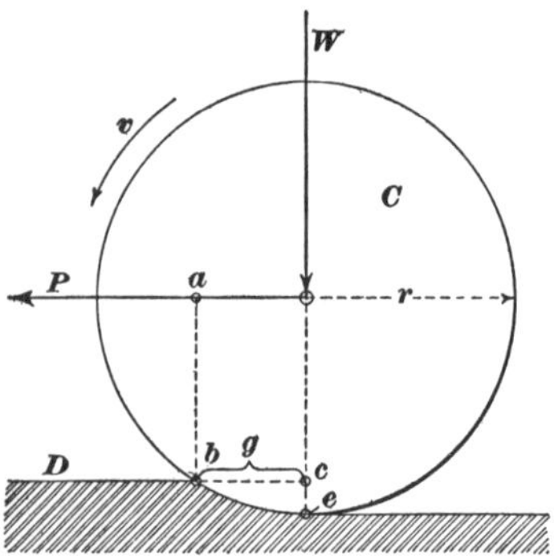

FIG. 24

$$P \times a\,b = W \times b\,c, \text{ and } P = \frac{W \times b\,c}{a\,b}$$

If $b\,c = g$ and the cylinder radius is r, then

$$P = \frac{Wg}{r - ce}$$

For hard materials, as used in bearings, the length ce is very small compared with r, and may be left out of consideration. The formula then becomes

$$P = \frac{Wg}{r}, \text{ approximately}$$

This formula shows that the rolling resistance decreases with an increase of r.

Very few experiments have been made to determine the value of g. Some tests give g a value of from .007 to .015 inch for steel wheels rolling on steel rails.

As the rolling resistance between two polished metal surfaces is very low, the advantage to be derived from this combination has been utilized in the construction of roller and ball bearings. In all bearings of this class, it is important that the rollers or balls, as well as the paths provided for them, be of the greatest hardness and smoothness. For this reason, if the whole of the parts in contact cannot be made of the required hardness, it is customary to provide them with linings that possess the requisite qualities.

35. Cylindrical Rollers.—The principal parts of a roller bearing, with the accessory parts omitted, are shown in Fig. 25, (a) being a longitudinal section and (b) an end

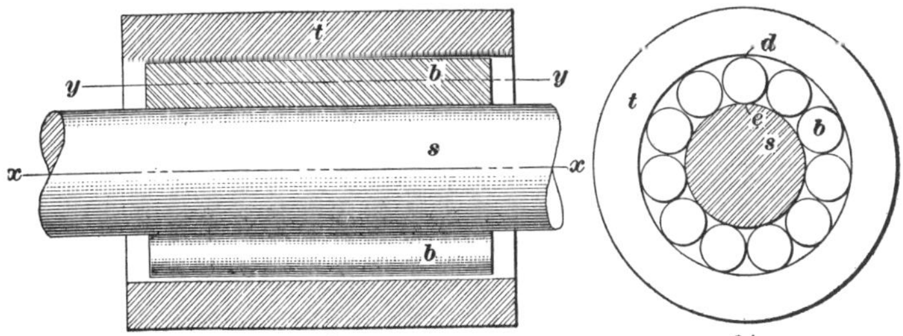

(a) (b)

FIG. 25

view. The part s may represent part of a shaft or journal surrounded by a series of rollers b, b, which are held in position by means of the box t. In order that pure rolling may take place, the casing, rollers, and journal must be perfectly cylindrical; also, the axes yy of the rollers and the axis xx

of the journal must be parallel, and endwise motion of the rollers must be prevented. For the purpose of holding the rollers parallel to one another and to the journal, various methods are in use, as slotted cylinders, or cages, rings with pins, and rods guiding and separating the individual rollers.

Rollers may be conical as well as cylindrical. In order that true rolling may be insured with conical rollers, it is necessary that all the axes of rotation of the rollers shall intersect the axis of the journal at the same point, the position of which remains constant throughout a revolution of the journal.

The application of this rule may be more clearly shown by means of Fig. 26, where the rollers *b*, as well as the journal *s*, are conical. The axes *y y* of the rollers, produced, intersect

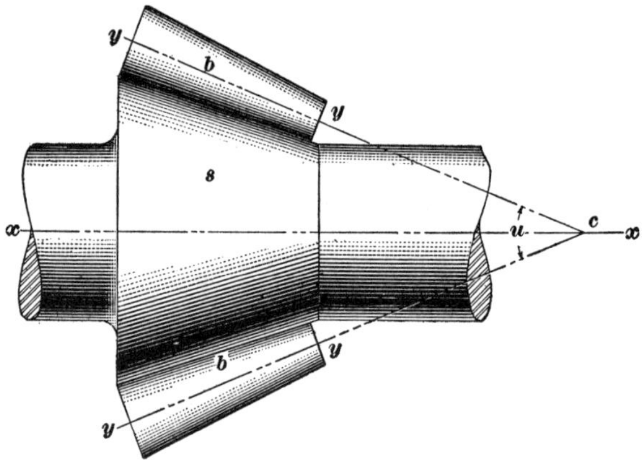

FIG. 26

the axis *x x* at a point *c*. If the position of this point is unaltered while the journal completes a revolution, true rolling takes place.

36. According to the preceding rule, there are cases where true rolling should occur, but where, nevertheless, sliding friction is present. An instance of this kind is shown in Fig. 27, which may be considered as a variation of Fig. 26, the angle u being increased to 180° so that the axes $y_1 y_1$ and $y_2 y_2$ are at right angles to $x x$. In Fig. 27, b_1 and b_2 are cylindrical rollers resting on the bearing plate *t* and supporting the circular end of the shaft *s*. If the rollers are guided by suitable means, such as a perforated disk, the intersection of the axes $y_1 y_1$ and $y_2 y_2$ with the axis $x x$ of the shaft will be

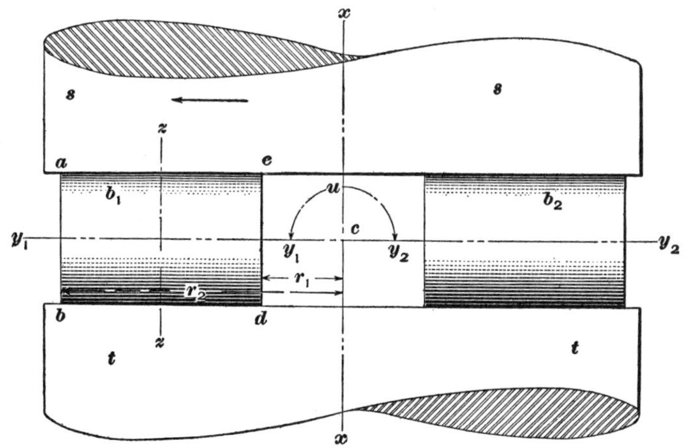

FIG. 27

at the point *c*, and the position of this point will be unaltered throughout a revolution of the shaft.

Owing to the difference in peripheral speed between the points *a* and *e* at the end of shaft *s*, it is evident that the end *a b* of the roller b_1 tends to rotate at a greater speed than the end *d e*. As this is impossible, it follows that there must be relative motion between one end or both ends and the contact surfaces, which will result in friction. As the rollers are caged, they are compelled to revolve around the axis $x x$ and, simultaneously, to revolve around a vertical axis located anywhere between *a* and *e*, as at $z z$. This latter motion is termed *spinning*, and is one that must be eliminated as far as possible if the bearing is to be efficient and durable.

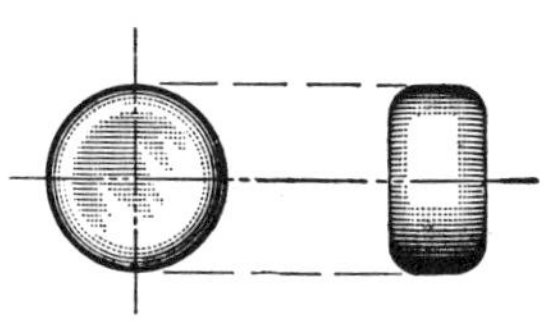

FIG. 28

Attempts have been made to obviate the injurious effect of spinning in thrust bearings, as in Fig. 27, by using a great number of very short rollers having rounded edges, as shown in Fig. 28. The difference between the circumferential speeds at the two ends of the roller is then decreased to such an extent as to be of little consequence.

Fig. 29 shows such rollers inserted in a *cage*, or guiding plate *e*, of a thrust bearing. This cage is provided with rectangular openings and several series of rollers, as a_1, a_2, etc., and b_1, b_2, etc., every alternate series being so arranged relatively to the others that the paths of any two series do not coincide; consequently, the wear is more evenly distributed. The axes of all the series converge toward the common center *c* located on the axis of the shaft. The cage *e* rests on a raised edge, or flange, *d*.

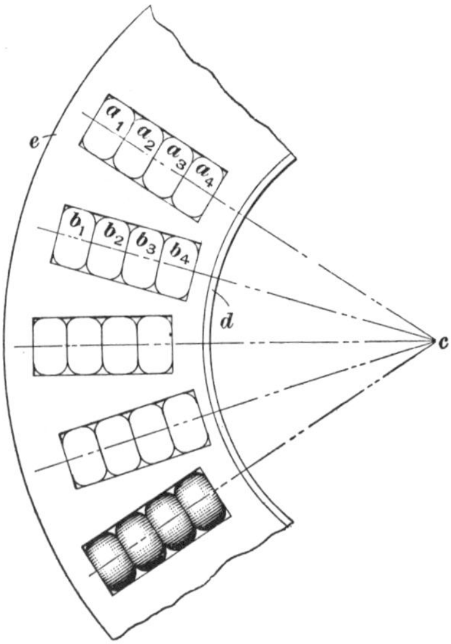

Fig. 29

Under certain conditions, thrust bearings of this kind have been very successful and have sustained very high pressures; both the rollers and the contact plates were made of hardened tool steel. Great care should be taken to see that the diameters of the rollers are perfectly uniform, and that the surfaces are absolutely plane and highly polished.

37. Conical Thrust Rollers.—If rollers having their axes at right angles to the axis of the shaft are to comply with the rules for true rolling, they must be made conical, and be proportioned according to the principles illustrated in Fig. 30 (*a*). That is, any roller constituting the frustum of a cone will have a true rolling motion if its apex coincides with the apexes of the conical surfaces on which it rolls.

The apex angles u_1, u_2 of these surfaces do not have to be equal; they may be varied until one of them becomes 180°, as at u_2, Fig. 30 (*b*), and may be increased beyond 180° if necessary.

It can be proved that if the roller b_1 is to have a true roll-

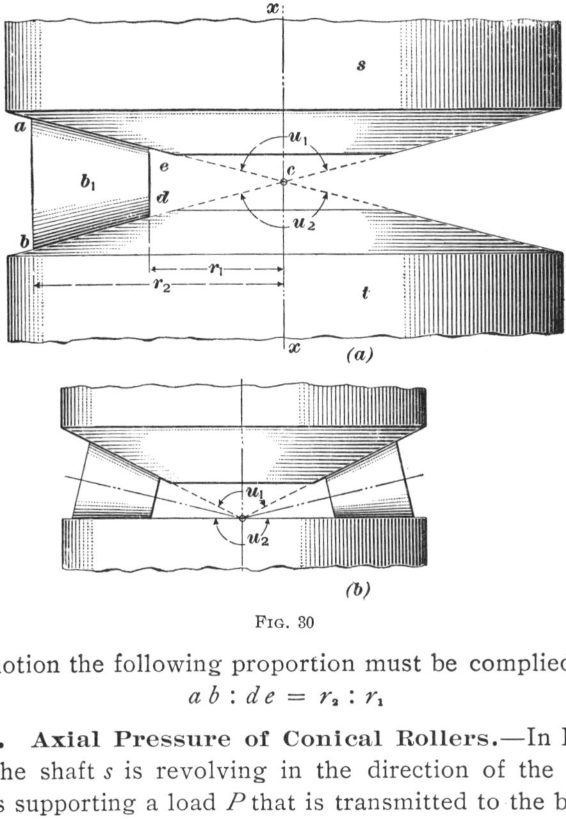

FIG. 30

ing motion the following proportion must be complied with:

$$a\,b : d\,e = r_2 : r_1$$

38. Axial Pressure of Conical Rollers.—In Fig. 31 (*a*), the shaft *s* is revolving in the direction of the arrow, and is supporting a load *P* that is transmitted to the bearing plate *t* by a series of rollers, two of which, *b*, b_1, are shown. Two forces P_1 and P_2 will therefore be acting normally to the sides of each roller. A resultant of these forces will act along the axis *c e*, tending to force the roller outwards; that is, away from the center *c*. Thus, a force P_3 will have to be applied at *e*, so as to prevent a longitudinal motion of the roller.

The magnitude of these forces may be ascertained as indicated in Fig. 31 (*b*). Draw the line *fg* parallel to *P*. Let P_0 represent the portion of the total load carried by each roller, and to some given scale, make *fg* of a length representing the magnitude of P_0. Draw *fh* at right angles to *fg*, and from point *g* line *gh* parallel to P_1. Then, the point of intersection *h* will determine the length of $gh = P_1$. As the action of P_1 is opposed by the reactive force P_2, draw

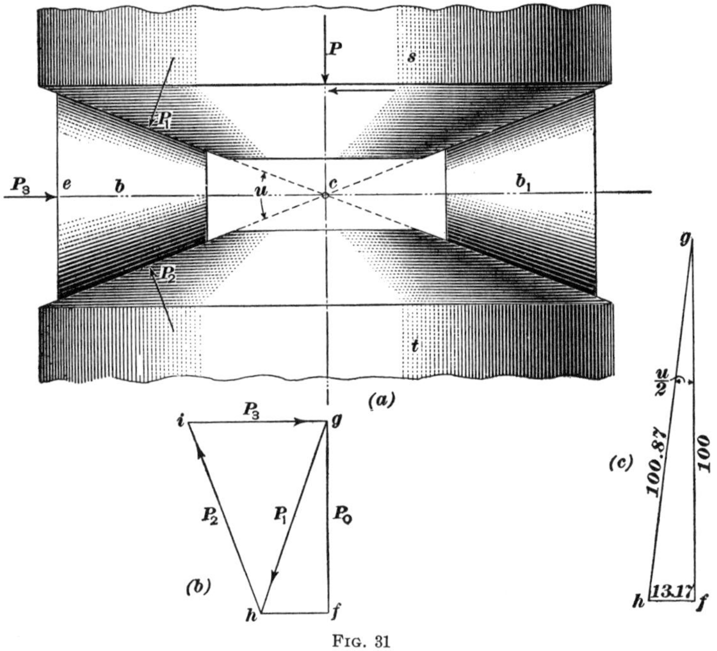

FIG. 31

the line *hi* parallel to the latter, and from *g*, the line *gi* parallel to the axis *ce*. Then, *gi* will indicate the magnitude of the reaction P_3 required to hold the roller in place. It is evident that this force P_3 should be as small as possible, and that, therefore, the angle *u* should be as small as the conditions will allow. This angle should not exceed 15°.

EXAMPLE.—In a roller thrust bearing similar to Fig. 31 (*a*), the pressure P_0 on each roller is 100 pounds. If the vertex angle *u* of the roller is 15°, what will be the pressures P_1 and P_3?

SOLUTION.—In solving the problem graphically, it is sufficient to draw the portion corresponding to fgh, Fig. 31 (*b*). Draw a line fg, Fig. 31 (*c*), representing the pressure of 100 pounds to any convenient scale, as, for instance, 1 in. = 10 lb. Therefore, fg = 10 in. Lay off the angle $fgh = \frac{1}{2}u = 7° 30'$, and draw gh; then, draw fh perpendicular to fg. By measuring the lines according to the chosen scale, it is found that the line $fh = \frac{1}{2}P_3$ = 13.17 lb., and, therefore, P_3 = 26.34 lb. The normal pressure $P_1 = gh$ = 100.86 lb. Ans.

An example of a roller thrust bearing is shown in Fig. 32, which illustrates a full-sized transverse section through one side of the bearing. The bearing plate *g* revolves with the

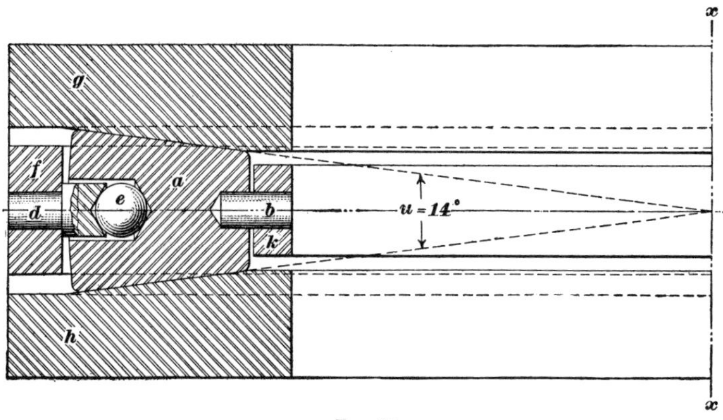

FIG. 32

shaft, and the plate *h* is stationary. One of the conical rollers is shown at *a*, and is held in position by the pins *b* and *d*. A ball *e* is inserted in a recess in the roller and rests against the concave end of the pin *d*, its purpose being to reduce the frictional resistance between the roller and the pin *d*. The inside pins are supported by the ring *k*, and the outside ones by the ring *f*, both rings revolving around the axis xx. In this instance, the apex angle *u* is 14°.

39. Most of the roller bearings manufactured come under one of the classes just described. On account of the rigidity of the parts, all bearings must be accurately shaped, and the material used must be extremely hard. In one form of roller bearing, the rollers are not solid, but are made of steel bands wound spirally into the shape of a roller. Rollers constructed in this manner have a certain amount of flexibility, and thus require less accuracy in their manufacture; likewise, the surfaces against which the rollers revolve need not be of such extreme hardness.

40. Construction of Roller Bearings.—The strength of a roller bearing cannot be safely determined except by actual test; it depends to a great extent on the material used for the rollers and bearing surfaces. In some cases, bearings that are identical in size and construction will show great differences in durability owing to insufficient hardening of some of the parts. Calculations based on the strength of the material and the form of the rollers are not safe guides, as the actual strength is often found to be far below the estimated value. Extensive tests have been made by Striebeck, the results of which are embodied in the following formulas:

Let P = total load on bearing, in pounds;
P_1 = safe load for one roller;
l = length of roller, in inches;
d = diameter of roller, if cylindrical, and average diameter, if conical;
k = a coefficient which is 1,000 for hardened steel;
n = number of rollers in bearing.

Then, for a thrust bearing in which the load is supposed to be evenly distributed among the rollers,

$$P_1 = \frac{P}{n} \qquad (1)$$

For a cylindrical bearing in which the load acts in a direction normal to the journal axis, some of the rollers will carry the load while the rest are idle. Under these conditions, it is generally found that

$$P_1 = \frac{5P}{n} \qquad (2)$$

The safe load for one roller in either of these bearings is

$$P_1 = kdl \qquad (3)$$

41. Before constructing a roller bearing, it is necessary to ascertain the number of rollers required to support a given load. Having found this number, the diameters of the surfaces along which the rollers revolve must be calculated. Referring to Fig. 33, let

D_1 = diameter of circle on which the roller centers are located;

D_2 = diameter of circle tangential to the inner side of the rollers;

d = diameter of rollers;

C = diameter of circle circumscribing the rollers.

Then,
$$C = D_1 + d \qquad (1)$$
and
$$D_2 = D_1 - d \qquad (2)$$

From the center o, draw radii $o\,l$ and $o\,m$ to the centers of two adjoining rollers. Connect these centers with the chord $l\,m$, and draw a line $o\,n$ to the point of tangency between the two rollers. Then, the angle $l\,o\,m = \frac{360}{n} = 2\,u$, and the angle $m\,o\,n$ is

$$u = \frac{180}{n} \qquad (3)$$

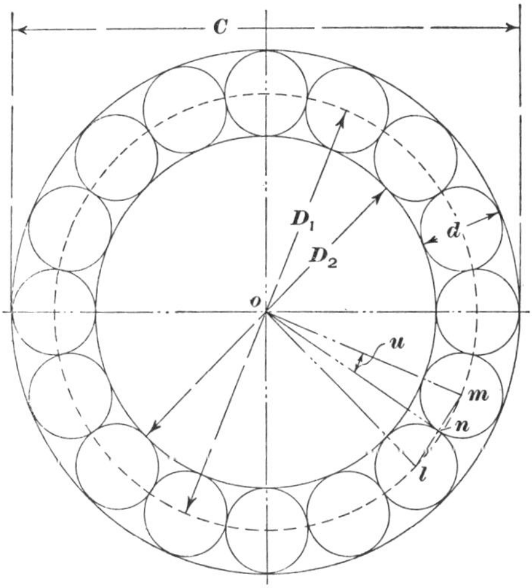

FIG. 33

In the right triangle $m\,n\,o$, the side $m\,n = m\,o \times \sin u$; and as $m\,o = \frac{1}{2}D_1$ and $m\,n = \frac{1}{2}d$, it follows that $\frac{1}{2}d = \frac{1}{2}D_1 \sin u$, or $d = D_1 \sin u$. Hence,

$$D_1 = \frac{d}{\sin u} \qquad (4)$$

Having found the value of D_1 by means of formula **4**, the values of C and D_2 may be found from formulas **1** and **2**.

These formulas are based on the assumption that all the

rollers are in contact with one another. In practice, the cage in which the rollers are inserted leaves a certain amount of space between each pair of rollers, and this space must be considered in using formula **4.**

Let i = space between two adjoining rollers;

d_1 = roller diameter plus one-half the space on both sides (which is equal to the space between two adjoining rollers), or $d + i$.

Then, $\frac{1}{2} d_1 = \frac{1}{2} D_1 \sin u$; $d_1 = D_1 \sin u$

and
$$D_1 = \frac{d_1}{\sin u} \qquad (5)$$

The value of the angle u, as found from formula **3,** is not affected by the addition of the space i. The values of C and D_2 for the new value of D_1 are still found by means of formulas **1** and **2.**

EXAMPLE.—A roller bearing is to contain twelve cylindrical rollers, each 1 inch in diameter. Assuming that there is to be a space of $\frac{1}{4}$ inch between each pair of rollers, find the diameter of the circle passing through the roller centers of the journals, and of the outside casing.

SOLUTION.—From formula **3,**
$$u = \tfrac{180}{12} = 15°, \text{ and } d_1 = d + i = 1 + .25 = 1.25 \text{ in.}$$
From formula **5,**
$$D_1 = \frac{1.25}{\sin 15°} = \frac{1.25}{.2588} = 4.83 \text{ in.}$$
From formula **1,**
$$C = D_1 + d = 4.83 + 1 = 5.83 \text{ in.}$$
From formula **2,**
$$D_2 = D_1 - d = 4.83 - 1 = 3.83 \text{ in.} \quad \textbf{Ans.}$$

BALL BEARINGS

42. Distinctive Features.—Fig. 34 (*a*) illustrates a cylindrical roller *a b c d*, having hemispherical ends. If the length of this roller is reduced until the lines *a b* and *c d* coincide, as in (*b*), a ball is produced. Hence, **ball bearings** may be considered as a special form of roller bearings in which the length of the linear contact is reduced to a point. In reality, owing to the elasticity of the material, the contact of a roller with its support does not constitute a line, but rather a rectangular surface. Similarly, the contact between a ball and its support is not a point, but a circular or an elliptical surface, depending on the shape of the support. In all theoretical considerations, however, it is assumed that there is either line or point contact.

In reducing the length of a roller to that of a ball, certain advantages are gained. By reason of the line contact of the rollers in a roller bearing, they suffer from a certain lack of adjustability and from stringent requirements as regards the exactness of the engaging surfaces and the parallelism of their axes. If the journal should show signs of deflection under load, it would materially affect the efficiency and durability of the bearing. In a ball bearing, the length of the line contact is reduced to a minimum; hence, the adjustability of the rolling parts is a maximum. If, in roller bearings, it is important that true rolling take place, and that material and finish be of the best, it is still more so in ball bearings, where the supporting area is more limited and therefore more sensitive to irregularities in relative motion and character of the contact surfaces.

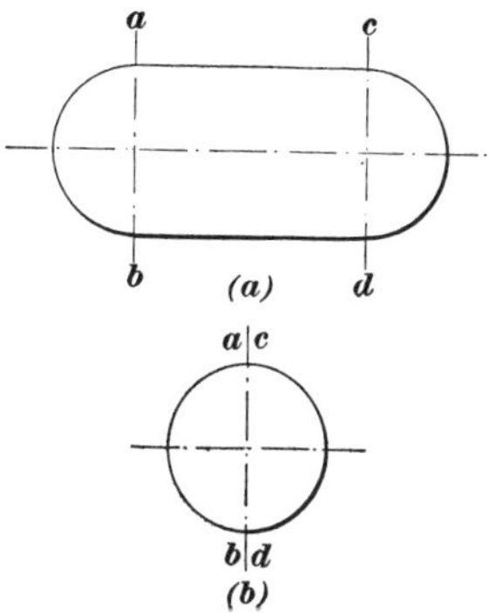

Fig. 34

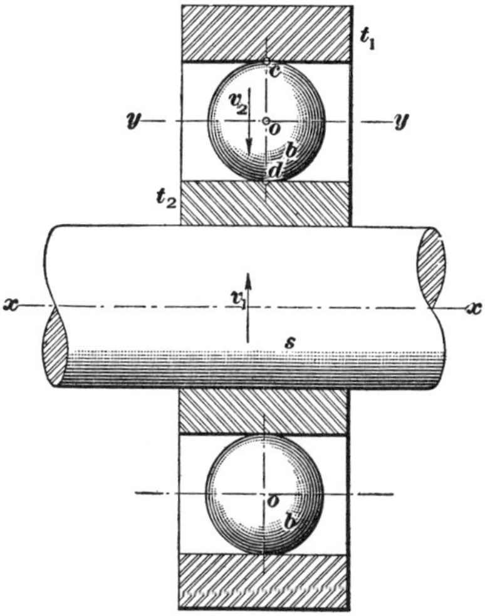

Fig. 35

43. Radial Bearings.—The fundamental form of a ball bearing is one in which the axes of rotation of the balls are parallel with that of the journal, as shown in Fig. 35, in which s is the journal, t_1 the *outer race*, t_2 the *inner race*, and b, b the balls. The axis xx of the journal is parallel with the axis of rotation yy of any one ball. This bearing corresponds in principle to the roller bearing shown in Fig. 25, and is termed a **radial ball bearing,** because the two points of contact c and d lie on a line radial to the journal axis.

44. As a bearing of the form shown in Fig. 35 would not, without the use of supplementary means, prevent the balls from leaving their races, the latter are provided with grooves c_1 and c_2, as shown in Fig. 36. In cross-section, these grooves appear as circular arcs of radii r_1 and r_2. which are somewhat larger than the radius of a ball. Authorities differ as to the most suitable ratio between these radii, but Striebeck, from extensive experiments, recommends $r_1 = r_2 = \frac{2}{3} d$, where d is the diameter of a ball. These proportions are given the grooves shown in Fig. 36.

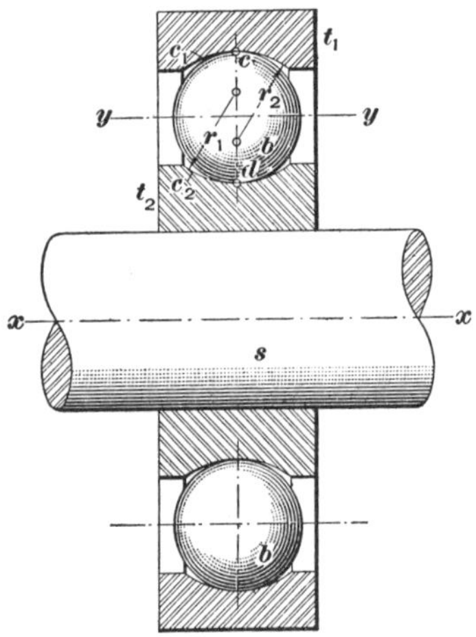

FIG. 36

In a bearing of this form, the journal is still allowed a certain amount of axial play by reason of the flatness of the curves in the grooves. This form possesses practically the same quality of pure rolling inherent in the form shown in Fig. 35, but has in addition the advantage of a greater surface contact, because the curvature of the supporting surface conforms more nearly to that of ball. Ball bearings of this class have proved very efficient, even under heavy pressures. If the curvature of the ball race is made as shown in Fig. 36, the end thrust of the journal may be as high as one-third of the load, normal to the journal axis, without requiring any provisions for preventing end motion.

45. Another variation of Fig. 35 is shown in Fig. 37 (*a*), in which one of the rolling surfaces, instead of being cylindrical, consists of two conical surfaces joined along the line *fg*. This variation is termed a **three-point bearing,** there being three contact points *c*, *d*, and *e*, which are points of tangency between the ball and the ball races. In this example, the balls partake of another motion besides pure rolling. This motion may be seen more clearly in Fig. 37 (*b*) and (*c*), where the angle between the sides of the groove is made more acute. If while rolling along the race t_1 the ball *b* turns in the direction of the arrow v_1, then, at any moment, the ball revolves around a line *z z*, passing through the points *c* and *d*, as an instantaneous axis of rotation. Theoretically, these places of contact should be merely points, but, actually, they are two circular surfaces with *c* and *d*, respectively, as centers. By looking along the line *i o*, the ball will appear as in Fig. 37 (*c*), revolving around the center *c* in the direction of the arrow v_2. The contact area at *c* may appear somewhat like the shaded circular surface, having a sliding friction similar to that found at the end of a revolving shaft resting on a plane surface.

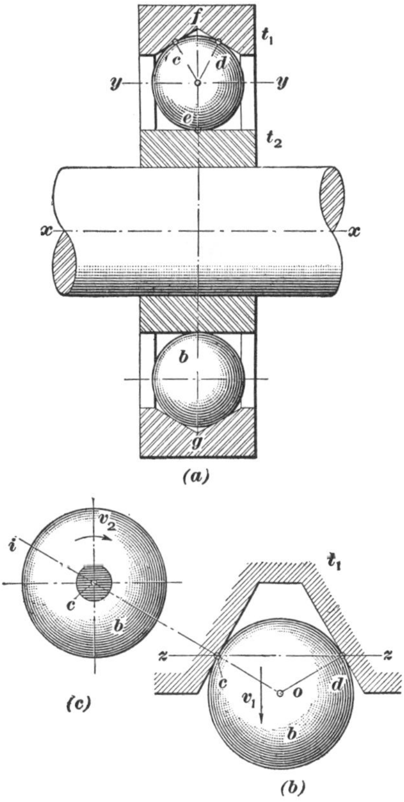

FIG. 37

This supplementary motion of the ball around a point of contact, termed spinning, should be absent from any bearing where high efficiency is required. The spinning motion of a ball is somewhat similar to that of a top, which revolves around its axis while it advances over the supporting surface.

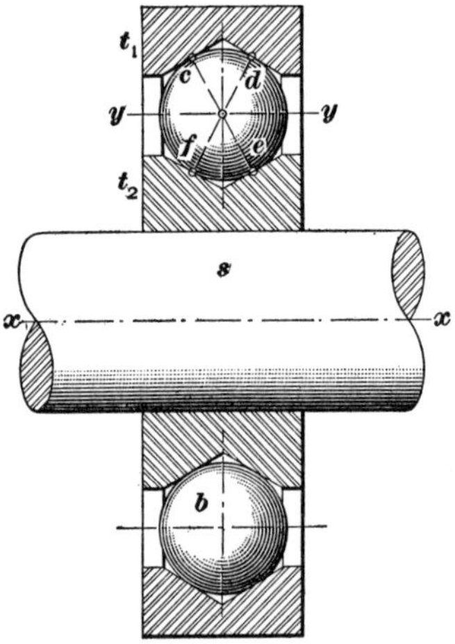

FIG. 38

46. If the race t_2 also is provided with a groove, as in Fig. 38, then a **four-point bearing** is produced, having contact point c, d, e, and f, at each of which spinning takes place.

47. Thrust Bearings.—In **thrust bearings**, the axis of rotation $y\,y$ of the ball is normal to the axis $x\,x$ of the journal, as shown in Fig. 39. This bearing corresponds to the roller bearing illustrated in Fig. 27. In this form, the balls will not retain their positions unaided, and a cage, or perforated plate, must therefore be used for the purpose. The balls while following a circular path with radius R will revolve, or spin, once around the axis $z\,z$ for each revolution around the axis $x\,x$.

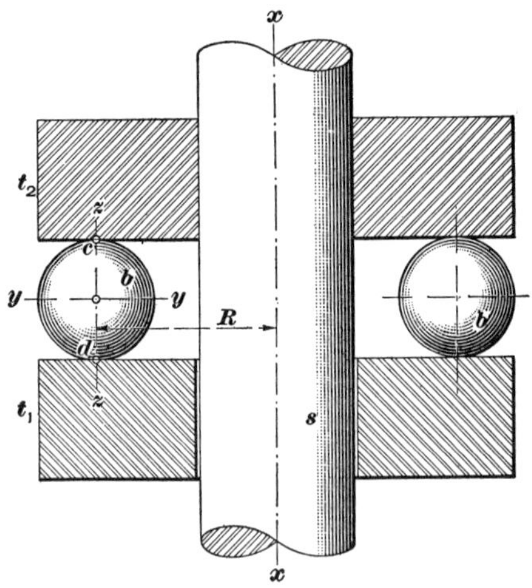

FIG. 39

48. To prevent the balls from running off tangentially to their circular path, one of the races, as shown in Fig. 40, is provided with a groove having sloping sides. As here shown, the ball will not have a true rolling motion for the following reasons: It may be considered part of two separate cones, one of which has its vertex at z_1 and rolls along the plane cz_1, while the other has its vertex at z_2 and rolls along the plane dez_2. Evidently it is impossible for two cones revolving around different axes to be rigidly connected through the ball b; hence, there must be a resultant of the two motions, represented by a common axis of rotation, having one point at o and another point somewhere between z_1 and z_2, as at z. There will be spinning at all three points of contact. If the race t_1 causes an excess of resistance to relative motion between it and the ball, then the spinning at c will be increased and the point z will be nearer z_2.

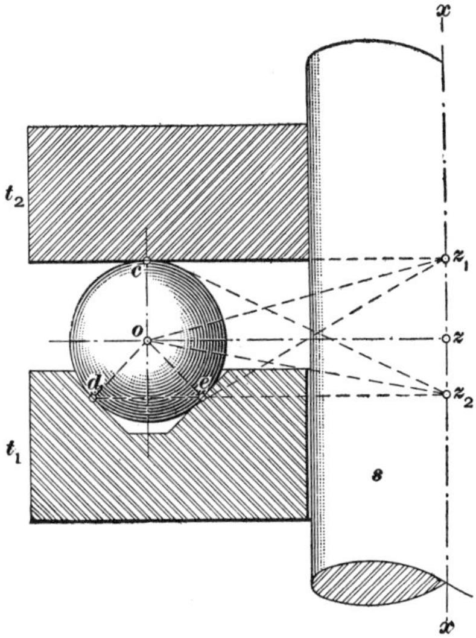

FIG. 40

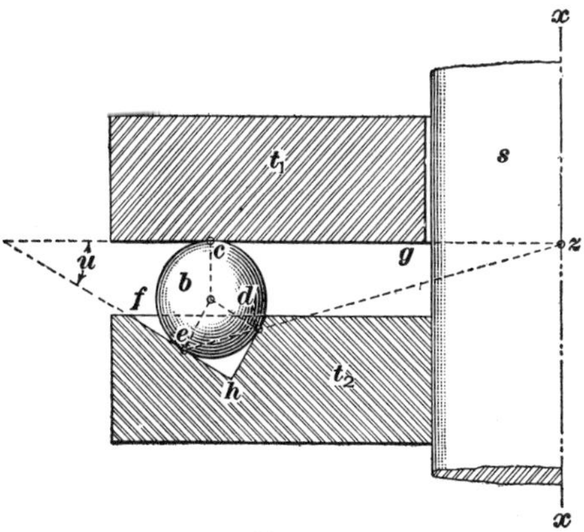

FIG. 41

49. Fig. 41 illustrates a rearrangement of Fig. 40, in compliance with the requirements for true rolling. The method of construction is as follows: If cg is to be one of the bearing surfaces, produce the line cg until it intersects the axis xx at z, which point will be the vertex of an imaginary cone. From z draw a line intersecting the ball b at two points, as d and e, which are to act as contact points between the ball and the race t_2. The sides fh and dh of the groove should be tangent to the ball at these points, as shown. While a certain freedom prevails in the choice of the contact points by varying the angle cze, there is a position below which the point e must not fall. Experiments have shown that if the angle between any two adjoining surfaces, such as the angle u between the planes ef and cg, are below a certain value, the ball tends to wedge itself between the surfaces. The angle should be at least 25°.

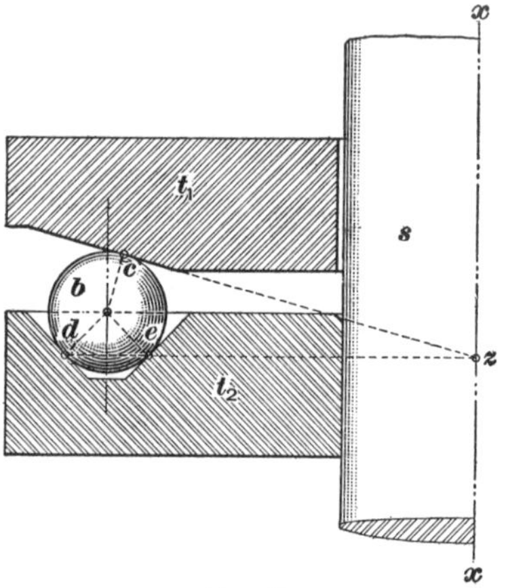

FIG. 42

On the other hand, the angle u should not be too large, as in that case the angle fhd would become too acute. This would give the ball a tendency to wedge itself in the groove at h, thereby greatly increasing the pressure at the points d and e where spinning takes place.

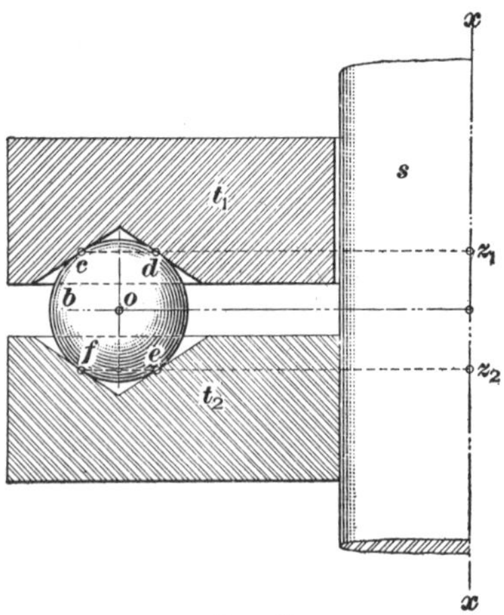

FIG. 43

50. The position of the vertex z may be changed without affecting the action of the ball. In Fig. 42, for instance, the point z has been lowered so as to make the lower side of the imaginary cone normal to the axes xx. If the construction is made as explained in Art. **49,** the conditions for practically true rolling are also here fulfilled.

51. Fig. 43 shows a form of ball bearing that has been extensively used in various machines, although it is very inefficient. In this instance, the lines $c z_1$ and $f z_2$, which intersect the ball at the points c, d, e, and f, are parallel. From the foregoing, it is clear that if the ball is supposed to be part of a cone, its sides should roll simultaneously along the surfaces $c z_1$ and $f z_2$; but as it cannot do so, there must be some sliding at one or more points, which should be avoided.

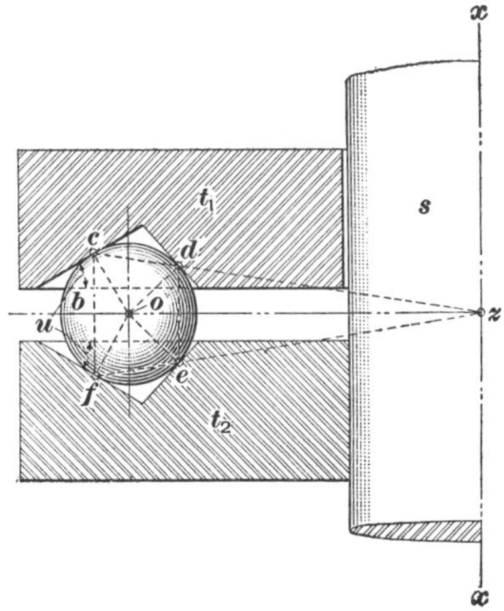

FIG. 44

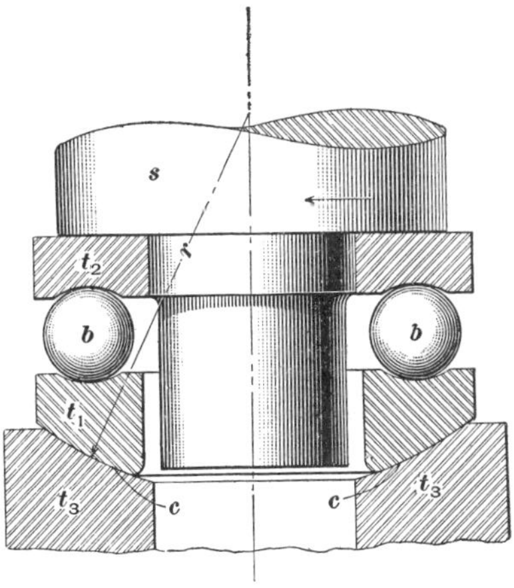

FIG. 45

52. In order to insure true rolling, the bearing should be constructed as shown in Fig. 44. When produced, the lines $c d$ and $e f$ passing through the four contact points should have a common point of intersection z located on the axis $x x$. Of course, the axis $o z$ need not be normal to $x x$; the point z can be moved to some extent above or below its present position, but care must be taken that neither of the angles u, u is too small. In the diagram, each angle is 30°.

53. Fig. 45 shows a thrust bearing with ball races of the form shown in Fig. 36. The surface of plate t_1 is curved at c to the radius r, thus providing a certain amount of adjustability relative to the main bearing t_3. This is necessary if the load is to be evenly distributed among all the balls b. The plate t_2 is fastened to the shaft s.

54. Hub Bearings.—There is still another class of ball bearings, known as **hub bearings,** which may be considered as a combination of a radial and a thrust bearing. This type differs from the regular thrust bearing in that the end thrust does not proceed from without the bearing, but

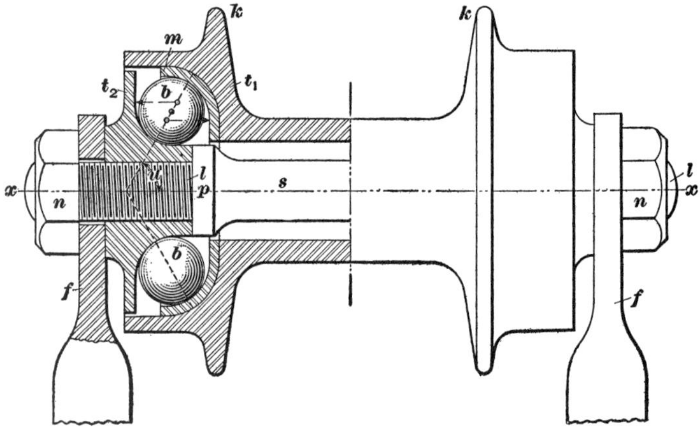

Fig. 46

from within, as a result of the arrangement of the ball races. These bearings are always used in pairs, so that the end thrust produced by one is neutralized by that produced by the other. The evolution of the bicycle served to bring this class of bearing into prominence.

In order that its action may be better understood, the type of hub bearing used in a bicycle wheel is shown in Fig. 46. The stationary spindle, or axle, s of a rear wheel has threaded ends l, l, which engage with nuts n, n and cones t_2 on each end. The cone shown is permanently screwed against the shoulder p; at the other end of s there is no shoulder, thus permitting the other cone to be moved longitudinally on l

in order that the bearings may be adjusted. By screwing the nuts *n* tightly against the rear forks *f*, *f* of the bicycle frame, they are forced against the cones, preventing them from turning and at the same time fastening the spindle to the fork. The balls *b*, *b* transmit the load supported by the spindle to the cups or ball races *m* inserted in each end of the hub t_1. The hub is provided with flanges *k*, *k*, to which are fastened the inner ends of the spokes, not shown, which connect the hub and the wheel rim.

55. Fig. 47 shows, diagrammatically, one end of a **two-point hub bearing,** t_1 being the hub, t_2 the cone, *s* the spindle, or axle, *b* the ball, and *c* and *d* the two contact points

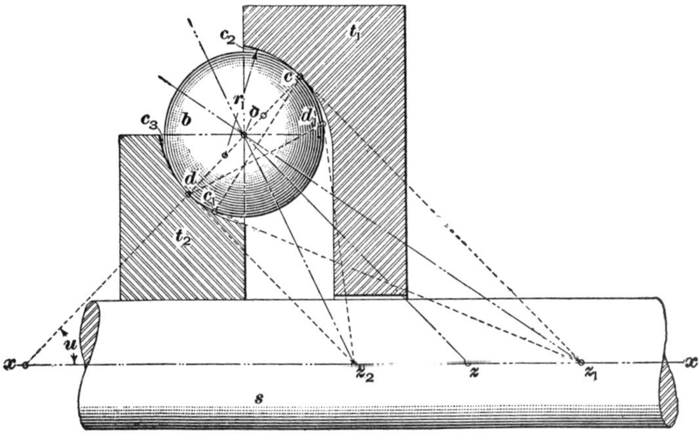

FIG. 47

with the races c_2 and c_3. The principles on which the action of this bearing is based are similar to those explained in Art. **48.** Drawing the tangents $c z_1$ and $d z_2$, and considering the ball as part of two cones with vertexes z_1 and z_2, respectively, it is again evident that true rolling cannot take place at both planes $c z_1$ and $d z_2$. In the present instance, the ball will revolve around an axis situated somewhere between $o z_1$ and $o z_2$, as, for example, at $o z$. Spinning will therefore occur at both *c* and *d*. In a bearing having the relative proportions shown in Fig. 47, it is found that for each revolution of the hub t_1 around the axis $x\,x$, the ball will make about four-fifths revolution around the axis $c\,d$. The amount of this rotary motion will increase with an increase of the angle u.

The radius r_1 varies between nine-sixteenths to three-fourths of the ball diameter. On account of this curvature the bearing cannot be so closely adjusted that the rolling paths of the ball will be strictly defined. For this reason, the hub has a certain amount of longitudinal play. This play will increase with an increase of the angle u, which, in a bicycle, may be as much as 60°, because there the side pressure is of secondary importance.

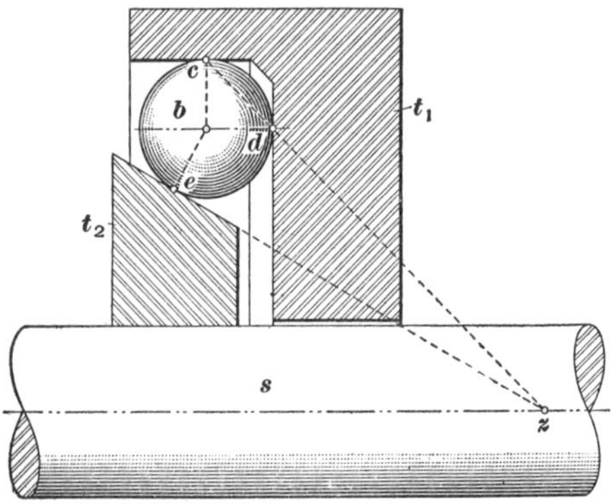

FIG. 48

56. Figs. 48 and 49 are diagrams illustrating **three-point** and **four-point bearings,** respectively. In Fig. 48, the contact points lie on the sides of the imaginary cone $c\,d\,z\,e$, and in Fig. 49, on the cone $c\,d\,z\,e\,f$. The method of construction is similar to that shown in Figs. 42 and 44.

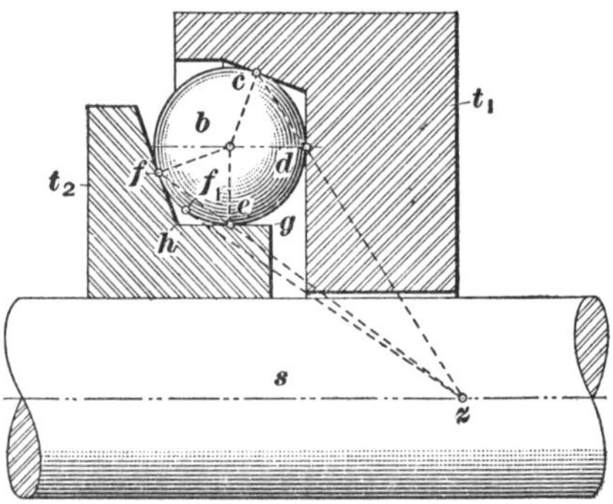

FIG. 49

57. By referring to the four-point bearings shown in Figs. 38, 44, and 49, it should be noted that these bearings require great accuracy in both their construction and adjustment in order to operate correctly. The reason for this may be seen, for instance, from Fig. 49. It is evident that if the cylindrical surface gh should have a radius somewhat smaller than shown, the ball would fail to make contact at e. The ball would still retain its position, but would make a three-point contact, the point f assuming the functions of the points e and f, but at a greatly increased pressure. It will be further observed that under these circumstances the position of point f is incorrect; it should occupy a position somewhere between points e and f, as at f_1, the point of tangency between a line $z f_1$ and the ball.

58. By comparing the two-, three-, and four-point bearings, it will be seen that in the last the balls are confined to a definite path, but that the bearing demands a very careful construction. In a two-point bearing, the path may vary on either bearing surface; hence, the bearing possesses a maximum of adjustability. A three-point bearing occupies an intermediate position, the rolling path being definitely determined for one surface, while that on the other has a certain amount of freedom.

59. Diameter of Ball Circle.—The diameter of the circle passing through all the centers o, Fig. 35, of the balls in a bearing is found, as described in Art. **41,** by means of formula **4.** This is under the supposition that the balls touch one another; but the conditions of easy running demand that a certain amount of clearance be left between the balls. The space left between each pair of balls varies with the total number of balls in the bearing. This space may be given a trial value of .005 inch, which is satisfactory if the sum of all the spaces does not exceed one-fourth of the ball diameter, otherwise the value should be reduced accordingly. The diameter of the ball circle is then found by means of formula **5,** Art. **41.**

EXAMPLE.—A ball bearing contains fourteen balls, each $\frac{1}{4}$ inch in diameter. Calculate the diameter of the ball circle if the balls are given suitable clearance.

SOLUTION.—If a clearance of .005 in. is given each pair of balls, the total clearance is $14 \times .005 = .07$ in.; as the value $\frac{d}{4} = \frac{.25}{4} = .0625$ in., the total clearance should be reduced to this amount. Therefore, the clearance for each pair of balls should be .0625 ÷ 14 = .0045 in.

From formula **5**, Art. **41**, $D_1 = \frac{d_1}{\sin u}$. Here, $d_1 = .25 + .0045 = .2545$ in. From formula **3**, Art. **41**, $u = \frac{180}{n} = \frac{180}{14} = 12° 51'$. Hence,

$$D_1 = \frac{.2545}{\sin 12° 51'} = \frac{.2545}{.22240} = 1.1443 \text{ in. Ans.}$$

60. Safe Load on Ball Bearings.—In thrust bearings of the type shown in Fig. 45, all the balls are supposed to take equal portions of the load, it being assumed that they are uniform in size. In bearings where the load acts in a direction normal to the axis of the journal, as in Figs. 36 and 46, the balls will be idle during one part of their travel around the journal and the load will be supported by two, three, or four balls, depending on the total number.

Let P = total load on bearing, in pounds;
P_1 = safe load for one ball;
n = number of balls in bearing, varying from ten to twenty;
d = diameter of the balls, in inches;
k = a coefficient that varies with the form of the bearing and the number of contact points. For hardened steel it has the following values: 400 for a straight-line race, as in Fig. 35; 700 for three- and four-point bearings; 1,400 for a two-point radial bearing, as in Fig. 36.

Then,
$$P_1 = k d^2 \qquad (1)$$

For a thrust bearing,
$$P = n P_1 \qquad (2)$$

For a bearing in which the load acts normal to its axis,
$$P_1 = \frac{5P}{n} \qquad (3)$$

For the radial type of bearings, as in Fig. 36, the load is independent of the speed if the latter is uniform and below 3,000 revolutions per minute. If sudden changes of load and speed take place, then the coefficient k should be lowered.

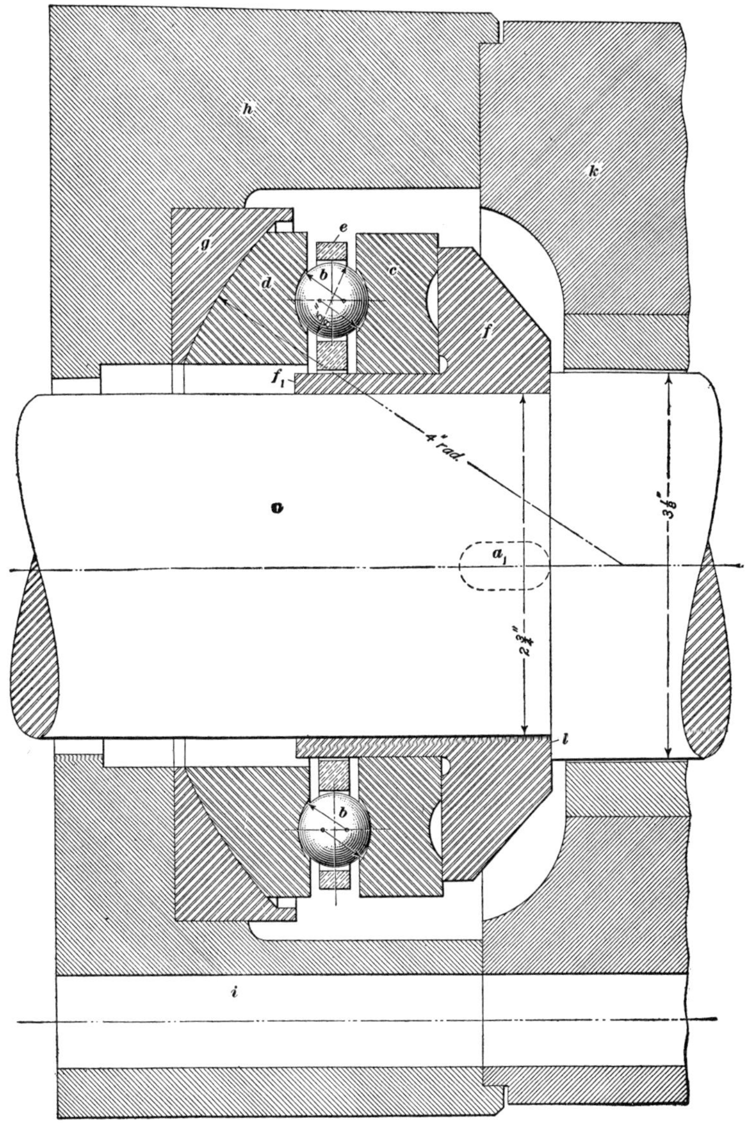

Fig. 50

EXAMPLE.—In a two-point ball bearing, as in Fig. 36, with fourteen balls each $\frac{1}{4}$ inch in diameter, the load is normal to the journal. Find the safe load.

SOLUTION.—From formula **1**,

$$P_1 = 1{,}400 \times .25^2 = 87.5 \text{ lb.}$$

From formula **3**, solving for P,

$$P = \frac{14 \times 87.5}{5} = 245 \text{ lb. Ans.}$$

61. Examples of Ball Bearings.—Figs. 50 and 51 illustrate two types of ball bearings. They are not given as standards of design, but simply as examples of bearings that have worked efficiently under the conditions for which they were intended. Both are two-point bearings with grooves of the form shown in Fig. 36.

Fig. 50 shows a longitudinal section of one of the thrust bearings of a mine pump that was required to be in constant operation, the bearing sustaining an end thrust of about 2,000 pounds and the shaft revolving at a speed of 890 revolutions per minute. A pump with this style of bearing ran continuously for 289 days.

In the illustration, a is the pump shaft, which is driven by an electric motor, and b, b the $\frac{5}{8}$-inch balls, of which there are sixteen, placed in holes in the cage e. The two halves of the ball bearing are lettered c and d. In its earlier form the ring c rested directly on the shaft, but it was found that the intensity of pressure at the shoulder l was too great, resulting in a partial crushing of the material and a tendency of the ring to tip—hence, the insertion of the bushing f with the extension f_1, and the key a_1. The end surface of the other ring d is made spherical to allow it a certain amount of adjustment; it rests against the concave side of the ring g. The casing h of the bearing is bolted to the main body k of the pump by bolts inserted through holes i.

62. In Fig. 51 is shown a pedestal bearing with fifteen balls, each $\frac{7}{8}$ inch in diameter, supporting a shaft $2\frac{9}{16}$ inches in diameter; (a) is a longitudinal section and (b), a side view, partly in section, the section being taken along the line $x\,x$ in (a). In this bearing, the balls b are not evenly

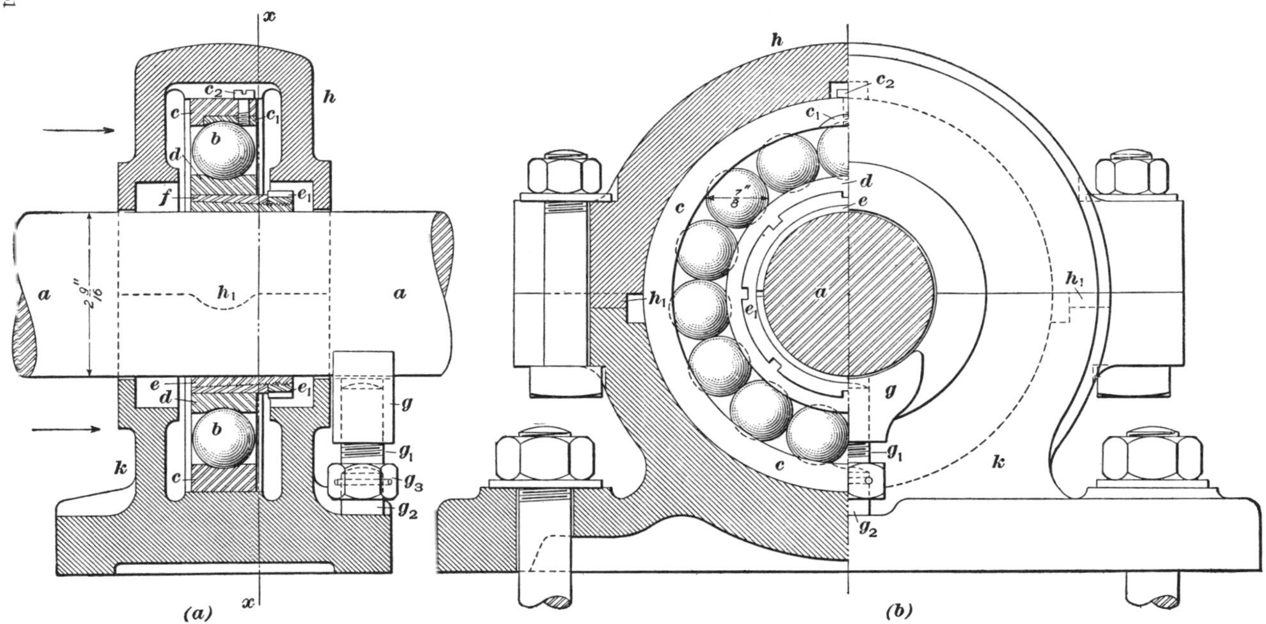
x
h
c
c_2
c_1
b
d
f
e_1
a
$2\frac{9}{16}''$
h_1
a
e
d
e_1
g
b
g_1
k
c
g_3
g_2
x
(a)
h
c_2
c_1
d
e
c
$\frac{7}{8}''$
a
h_1
h_1
e_1
g
g_1
c
k
g_2
(b)

FIG. 51

loaded, being idle while passing over the upper side of the shaft. The ball races are shown at *c* and *d*, the latter resting directly on the shaft if there is little vibration present. Otherwise, this race is fastened in the manner shown in (*a*). The sleeve *f* has a conical hole into which is fitted the split conical bushing *e*, one end of which is threaded to engage with the ring e_1. By turning up ring e_1, the bushing is drawn outwards, locking the ring securely to the shaft.

As rings *c* and *d* are not split, some means must be provided for the insertion or removal of the balls. For this purpose, a recess large enough to allow a ball to pass is made in the ring *c*. Into this recess is fitted a piece c_1, which is held by the screw c_2. The recess should be located at a place where the piece c_1 will not be subjected to any pressure from the balls. The outer casing is divided into two parts *h* and *k*, and is held together by bolts as indicated, the tongues h_1 insuring the correct alinement of the two parts. Between the casing *h* and the ring *c* there is a play of .004 inch, to allow for longitudinal adjustment of the rings and for expansion in case the parts become warm.

The chair *g* serves the purpose of supporting the shaft while the bearing is being inspected. The threaded end of the bolt g_1 is screwed into the chair, while the head revolves loosely on the stud g_2, to which it is connected by a pin, as indicated. By turning the head g_3, the chair may be raised until it supports the shaft, thus permitting the removal of the ball races.

In a more recent form of this bearing, the piece c_1 is omitted and the ring *c* is made solid throughout. To permit the insertion of the balls, their number is decreased, and they are inserted while the ring *c* is placed eccentrically relative to *d*. After the ball races are arranged concentrically, elastic distance pieces are inserted between each pair of balls, thus spacing them evenly around the shaft.

It is important that both roller and ball bearings be incased in such a manner as to retain a lubricant and exclude grit and dust. The lubricant will also prevent rust, but it should be free from acid.